FRONTIERS OF SCIENCE

This intriguing series encompasses exciting trends and discoveries in areas of human exploration and progress: astronomy, anthropology, biology, physics, geology, medicine, health, genetics, and evolution. Sometimes controversial, these timely volumes present stimulating new points of view about our universe . . . and ourselves. Among the titles:

Ancestral Voices: Language and the Evolution of Human Consciousness
Curtis G. Smith

The Human Body and Why It Works
Raymond L. Powis

Toward a New Brain: Evolution and the Human Mind
Stuart Litvak and A. Wayne Senzee

On Civilized Stars: The Search for Intelligent Life in Outer Space
Joseph F. Baugher

The Book of the Moon
Thomas A. Hockey

The LASER Book

The LASER Book

A NEW TECHNOLOGY OF LIGHT

Clifford L. Laurence

PRENTICE HALL PRESS · NEW YORK

To Bette,
who inspired me to reach higher

To the Life Training,
which showed me the way

Published by Prentice Hall Press
A Division of Simon & Schuster, Inc.
Gulf+Western Building
One Gulf+Western Plaza
New York, NY 10023

PRENTICE HALL PRESS is a trademark of Simon & Schuster, Inc.

Library of Congress Cataloging-in-Publication Data

Laurence, Clifford L.
  The laser book.
  (Frontiers of science)
  Bibliography: p.
  Includes index.
  1. Lasers—Popular works. I. Title. II. Series.
TA1680.L38 1986   621.36′6   86-8166
ISBN 0-13-523622-3

Designed by Jack Meserole/Publishing Synthesis, Ltd.

Manufactured in the United States of America

10 9 8 7 6 5 4 3 2 1

First Edition

Acknowledgments

The author wishes to extend sincere appreciation to Texe and Wanda Marrs, authors of many fine technical books, for serving as my agents and giving me the opportunity to write this book. Thanks to James Marlin Ebert for inspiring me to write the book and for introducing me to the Life Training. My gratitude to Phyllis Smalley for editing the manuscript with meticulous attention to detail. Many thanks to Dr. Mike Becker for his assistance in researching the book. And finally, my deepest appreciation to Dr. Frank Tittel, for introducing me to the fascinating fields of lasers and nonlinear optics, for serving as my professor and thesis adviser, and for his review of the manuscript and technical suggestions on the content of the book.

Contents

Foreword   xiii

Preface   xv

Introduction   1

1   What Are Lasers?   3

2   What Special Things Can Lasers Do?   6

3   Who Invented the Laser?   14

4   Things We Need to Know to Understand Lasers   22
     Scientific Notation   22
     Units of Measure   23
         Wavelength · Frequency · Linewidth (Bandwidth) ·
         Time Duration · Energy · Power
     Coherence   32
     Interference   34
     Index of Refraction   38
     Polarization   39
     Birefringence and Modulation   41

5   How Do Lasers Work?   43

Electron Energy Levels   43

Gain   47

Optical Resonators (Cavities)   49

Modes of Oscillation   52

Multimode and Single-mode operation   54

Q-Switching (Q-Spoiling)   54

Mode Locking (Phase Locking)   55

Modulation   58

Acousto-optic Modulation · Electro-optic Modulation

Summary of Laser Operation   60

Lasers That Don't Work the Way We Just Said They Do   61

Semiconductor Diode Lasers · Chemical Lasers ·
Gas-dynamic Lasers · Free-electron Lasers

6   Common Types of Lasers and
Their Characteristics   69

Helium-Neon (HeNe) Lasers   69

Carbon Dioxide (CO_2) Lasers   73

Sealed-tube CO_2 Lasers · Axial-flowing
Gas CO_2 Lasers · Transverse-flow CO_2 Lasers ·
Gas-dynamic CO_2 Lasers · Waveguide CO_2
Lasers · Transversely Excited Atmospheric (TEA)
CO_2 Lasers

Neodymium (Nd) Lasers   80

Dye Lasers   86

Excimer Lasers   90

Semiconductor Lasers   94

Gallium Arsenide Diode Lasers

Other Important Types of Lasers   99

Argon-ion Lasers · Helium-Cadmium (HECD)
Lasers · Copper Vapor Lasers (CVL) ·
Ruby Lasers · Vibronic Lasers · Nitrogen Lasers

Summary   104

7 A Myriad Laser Uses   106

    Applications in Research and Development   107

      Laser Spectroscopy · Energy Production
      Sources · Holography · Nonlinear Optics

    Applications in Engineering   125

      Optics and Precision Measurement ·
      Computer Engineering

    Industrial Applications of Lasers   137

      Materials Processing · Industrial Process Monitoring,
      Inspection, and Control

    Applications in Medicine   150

      Ophthalmology · Coronary and Arterial
      Medicine · Oncology · Neurosurgery · Gynecology ·
      Podiatry · Dermatology · Dentistry · Other Medical
      Applications

    Applications in Communications   166

      Fiber Optics · Atmospheric Optical Communications

    Military Applications   173

      Laser Designation · Tactical Laser Weapons ·
      Strategic Laser Weapons · Battlefield Simulations

    Lasers in Entertainment   183

    Summary   187

8 Education and Careers in Laser Technology   189

    In Conclusion   194

    References   197

    Index   201

    Credits   208

Foreword

Few developments have had more impact on as many different technical fields as the laser. The significance of this fact can be appreciated, understood, and fostered only by an informed public.

So often we look in awe at the technical achievements of our time without the means to understand and appreciate them. In this era of rapid development, technology is understood and appreciated more from the standpoint of what it can do for our lives and how much we can get for our money than for the potential value of underlying scientific principles or how these principles may contribute to other fields.

Two major problems persist. The first problem is how to keep young people informed so that they can understand technology sufficiently to make reliable career decisions. The second problem is how to keep people who specialize in one technology informed of the potential impact of developments in other technologies. Considerable effort is spent today to foster a more informed society by bringing technology to people beyond the scientific and technical communities. This book is a part of that effort.

Dr. Laurence's book attempts to strike a balance between the introduction of scientific concepts and the presentation of laser applications. The purpose is to keep all elements clear and concise so as not to discourage even the most casual reader. Only those concepts essential to an understanding of laser principles and uses are presented. Other books are available to fill in the details should the reader desire a greater depth of understanding in any given area. A broad sampling of applications is presented in fields in which lasers have had major impact. The coverage of applications is not intended to be comprehensive, but to give the reader a realistic feel for what lasers can do and how each unique feature of laser light can potentially be used.

This book constitutes a significant contribution in the effort to inform the public about technological advancements, to promote informed career decisions, and to enhance technical cross-fertilization.

—Dr. Frank K. Tittel
Quantum Electronics Laboratory
Rice University
Houston, Texas

Preface

This book is intended as a brief and concise introduction to the subject of lasers for people with nontechnical backgrounds. It may also be a valuable guide to lasers and their applications for technically oriented persons who have expertise in fields that are not laser related. The book may serve as a valuable source of information and a career guide for high school students.

A minimum number of technical concepts are required to understand what lasers are and how they work. Each concept is explained as carefully and as briefly as possible. A general but not in-depth knowledge of scientific concepts as learned in high school is assumed. Some of the concepts regarding lasers that are given here may be difficult for the nontechnical person and the high school student to grasp. However, an introduction to the concepts will promote easier understanding when they are encountered in the future.

The book discusses most major applications of lasers. Some are known to many people, but others may surprise readers. The total number of laser applications today is so staggering that all of them cannot possibly be covered. Professional laser magazines are recommended for keeping up with laser developments and the new applications that arise almost daily.

Introduction

T HE year 1960 stands out as a revolutionary year in the
world of science. It was then that scientists finally put
together all the necessary ingredients to create the first
laser. The year was revolutionary because the laser has developed into an incredibly useful device with an almost unbeliev-
able range of applications.

Lasers are sources of light. "Well," you say, "we have
plenty of sources of light. What's so special about lasers?" The
development of lasers is unique, and one of the most useful
technical achievements in history. First, the light is highly
single-colored; scientists use the term *monochromatic*. Second,
the light is highly organized; scientists use the term *coherent*.
Third, the light is highly directional; scientists use the term
collimated. Fourth, the light is available in a wide range of
power levels, from microwatts (millionths of watts) to terawatts
(million millions of watts). Fifth, the light is available continuously or in incredibly short, powerful pulses. And sixth, the
light is available in a wide range of colors (wavelengths from X
rays to microwaves).

In the coming chapters we will learn more about each of
these properties of lasers and how they fit together to permit
an immense variety of applications.

It is interesting to note that scientists knew everything that
was needed to build a laser as early as 1920, but no one put
together all the ingredients in just the right way until 1958.
Until then, there simply was no way to create a light source

1

with all the special properties of laser light. It should be emphasized that unless all the right ingredients are combined in just the right way, laser light cannot be generated. Furthermore, it is not easy to tell that you are even close to producing it. If any one element is not operating correctly, there is no laser light. As soon as all elements are correctly combined, POW!—the laser fires.

The basic elements required to make a laser are the lasing medium, some form of excitation source, and an optical resonator. The lasing medium is the substance that actually generates the light. It can be a gas, a liquid, or a solid. There are many substances that can be used as lasing mediums, but they must have certain properties, which we will learn about in chapter 1. The excitation source must supply energy to the lasing medium; scientists use the term *pumping*. One of the most fundamental principals of science is that you do not get something for nothing. We have to supply energy in order to get the laser's optical energy out. Pumping means that the excitation source changes the electrons that surround the atoms of the lasing medium from ground states to higher, excited energy states. These states are capable of emitting optical radiation. The optical resonator provides feedback into the lasing medium, stimulating the production of light. This permits the energy in the light beam to build up, or amplify, until a portion of it is transmitted by one of the optical resonator mirrors outside the laser. There it can be put to use.

As we shall see in chapter 7, the variety of applications for lasers is so broad that there are a very large number of avenues and directions that permit one to pursue a career using lasers or laser-related applications. The areas of academic interest most likely to involve the design or use of lasers are those related to physics, chemistry, electrical engineering, optics, and materials science. However, lasers are used in nearly every scientific and engineering discipline and in many associated technological applications. You do not necessarily have to become a physicist, an electrical engineer, or a Ph.D. to become involved with the fascinating world of lasers. We will explore these areas of study in chapter 8.

1

What Are Lasers?

A LASER is a device that uses a means of excitation to pump energy into the electrons of a lasing medium, which in turn emits and amplifies a light beam formed by mirrors at each end of the lasing medium. The mirror at one end is partially transmitting, which permits the amplified light beam to be emitted from the laser. A simplified schematic diagram of a laser is shown in figure 1.1. Depending on the type of laser, the lasing medium may be a solid, a liquid, or a gas. The excitation or pumping device is the laser's source of energy and may be electrical (such as a discharge), chemical, nuclear, or optical (such as a flash lamp or another laser). The mirrors at each end of the lasing medium form what scientists call an optical resonator or optical cavity. The optical resonator insures that the radiation emitted by the excited lasing medium is amplified until the losses are overcome and there is enough energy in the light beam to couple part of it out of the resonator.

Let us look a little more closely at the basic ingredients of a laser. The lasing medium is the substance that originates and amplifies the laser light. It is the heart of the laser. Scientists have identified thousands of substances that can be used as lasing mediums. As you might imagine, not just anything can be used as a lasing medium. What are the requirements? First, the substance must be transparent to the light that it produces. Otherwise, the optical beam cannot build up; it will be absorbed as quickly as it is produced. Second, the electrons that sur-

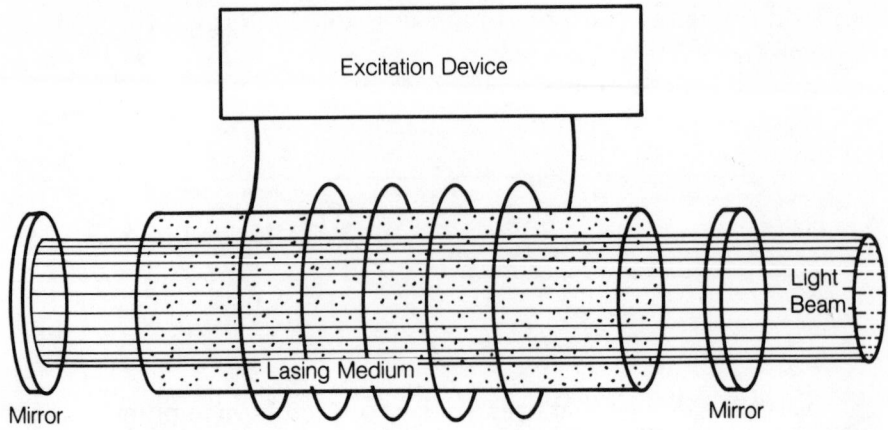

Figure 1.1. Schematic Diagram of a Laser

round the atoms of the lasing medium must have energy states that are quasi-stable. Let us talk for a moment about energy states.

All the atomic elements consist of atoms that are formed by a nucleus and surrounding electrons. The electrons can exist in many energy states but are usually in the lowest energy state allowed by nature. These states are stable; they will not change over a long period of time. However, they can be excited to higher states by the presence of outside energetic electrons or by pulses of light. When excited by absorbing energy from the outside electrons or light fields, the electrons of the lasing medium move to higher energy states. These higher energy states are usually not stable; this means that without excitation the electron immediately returns to a lower energy state. However, some excited states are quasi-stable, meaning that the electrons may stay in that energy state for a longer than usual period of time. These quasi-stable states are the essential property of the lasing medium. They make lasers possible. As electrons rise to higher energy states, they absorb energy, and when they fall to a lower energy state, they emit energy. The emitted energy is usually in the form of light. The energy states of electrons are not just of any energy level or value but are of a predetermined set of energy levels or values. This is why we refer to them as states.

When the states are quasi-stable, the electrons in the atoms of the medium that get into the upper or excited energy level are more numerous than the electrons that remain in the lower energy level. This phenomenon, called *population inversion,* permits the medium to emit more light than it absorbs. Many electrons of atoms are in the same state at the same time, another fundamental fact of nature that makes lasers possible.

As an electron moves from the excited state to the lower state, it emits a photon of light. Light occurs in very small energy bundles called photons. The color—or, as a scientist would say, wavelength—of the light is determined by the difference in the energy level of the excited state and that of the lower state. When population inversion occurs, it is possible for the excited electrons to amplify the light beam. We also know that when an electron is in this quasi-stable state and a photon of light passes nearby, the light may cause the electron to change states and emit another photon. This is called stimulated emission. Thus, a few electrons in the medium may spontaneously emit photons by changing states, and others will respond to these photons by changing states. In this manner the light beam is amplified until it builds up to a level where it removes energy from the lasing medium as fast as the excitation device can send more electrons into their excited state. Note that when a photon causes stimulated emission from an excited electron, the newly emitted photon is in nearly the same direction and is in step, or coordinated with, the stimulating photon. This is what causes the laser light to have the very important property called coherence. It is a unique and very useful property of light that so far cannot be produced in any way except with a laser. We will explain coherence in more detail in chapter 4.

The role of each primary ingredient of a laser is summarized here. The excitation device sends the electrons of the lasing medium into their quasi-stable excited energy states. The mirrors feed back the spontaneously emitted photons so that they can oscillate and be amplified by stimulated emission to form the laser beam. Now we can appreciate the meaning of the name *laser.* It stands for *l*ight *a*mplification by *s*timulated *e*mission of *r*adiation. Radiation is a general term for electromagnetic energy, which includes light.

What Special Things Can Lasers Do?

I n THIS chapter we will talk about the unique characteristics that laser light possesses and, because of these characteristics, some of the things that can be done with laser light. Literally thousands of applications have evolved from scientific research. They have created new or improved techniques used in industry, and have enabled medical science to do things that were previously impossible.

The applications that have been found for lasers range from the practical to the sublime, and unfortunately, in a few cases this includes the harmful. Lasers may soon become a source of energy concentrated enough to start thermonuclear reactions, which in turn can be controlled to produce unlimited energy for our use. Lasers used in medicine are saving lives and improving the quality of life for people almost every day. Lasers are even being used in systems designed to measure a baseball batter's performance so that he will know exactly what to work on for improvement. Lasers help make it faster and easier for you to check out at the supermarket. They can also be used as weapons.

To better understand the usefulness of lasers, let us summarize the special properties of laser light.

- It is intense; that is, the energy is concentrated, with output ranging in power from microwatts for small semiconductor lasers to billions of watts for large, pulsed, solid-state lasers, and output ranging in wavelength from the X-ray region to the microwave region.

(Photograph courtesy of AT&T Bell Laboratories, Murray Hill, New Jersey)

One of the smallest lasers made, the cleaved coupled-cavity diode laser, is shown among single grains of common household salt. This little laser holds significant promise of advance in optical communications through use with fiber-optic cables.

- It is highly directional (collimated), traveling in a narrow cone. This means it can travel over long distances while keeping its energy concentrated. The property of collimation permits laser light to be focused by a lens to a point where the concentration of energy is much larger than was ever before possible.
- It is monochromatic. This means it is highly single-colored. It can be used to selectively interact with materials, reading out information from preselected energy states in atoms and molecules.
- It has coherence. This means that it is highly organized. The waves that characterize the location of the light are organized so that they stay together (stay in phase) over relatively long distances and stay together over relatively long periods of time.

This is Pharos III, one of the largest lasers. This solid-state pulsed high-energy laser is being used at the Naval Research Laboratory for laser fusion research, high-altitude nuclear effects, and X-ray laser research.

- The energy in the light can be generated in extremely short pulses that result in remarkably high power levels.

Until the invention of the laser, there was no way known to science to generate light with these characteristics. This is why the invention of the laser was such an important advance for science and mankind.

One feature of certain types of lasers is a high degree of collimation. This means they can travel over long distances in straight lines without losing energy. Do you think you could see a light bulb from the moon? Certainly not, but you could see a laser, if it were properly directed toward you. As a matter of fact, an interesting application of lasers has been to measure the distance to the moon, and to measure it to within a few centimeters (about an inch)! Quite a feat and a very useful one that provides important information about the universe. Because of collimation, lasers can be used to measure distances (ranging from tens of meters to the distance to the moon), to aim rifles, to illuminate targets, to attack flying targets such as missiles or satellites, to communicate without wires or radio, to locate objects and measure their position, to guide bulldozers to cut perfectly level fields suitable for irrigation, and to measure the contaminants in a smokestack—to name just a few.

The next important property of laser light is its monochromaticity, or its single-coloredness. In chapter 4 we will discuss just how monochromatic laser light is when compared to other types of light. For now, let us just say that laser light is about 10 million times more monochromatic than the best conventional laboratory light sources. Another very important fact about lasers is that their full power is available within a single-colored beam, whereas with conventional sources the monochromatic light is selected by separating it from light of many colors, a process that allows very little power to remain. The monochromatic nature of laser light is of value to scientists who use the light to bombard atoms and molecules, and to observe the absorptions and emissions that result, activities that reveal a great deal about the structure and interactions of atoms and molecules. This type of science is called *spectroscopy.* A highly monochromatic laser has been used to measure the length of a rod to one part per billion, the most accurate measurement of length ever made. This capability is important for the maintenance of standards of measurement by the National Bureau of Standards. Lasers are now being developed to measure the position of the mechanisms that store and retrieve information on computer data-storage disks; in the future, it will be possible to store 100 times the amount of data on computer disks because of this greatly improved accuracy in measuring the position of the mechanisms. All of these types of measurement rely on the laser light being highly monochromatic.

Because laser light is collimated, it can be focused to very high energy concentrations. As a matter of fact it can be focused to a spot with a diameter nearly equal to the wavelength of the light. In addition, because the energy is monochromatic a lens will focus all the energy at the same focal point. An argon-ion gas laser that outputs 1 watt of green light focused by an ordinary lens would have a power density at the focus of about 16 kilowatts per square centimeter (103 kilowatts per square inch). This tremendous energy-concentrating power of lasers has led to more useful and interesting applications than has any other feature of lasers.

Focused high-power continuous lasers are being used to weld steel up to a few centimeters (about one inch) thick. Lower power levels can be used to make very thin, accurately controlled welds in precision parts. The welds are better than with conventional welding methods because the light can be more accurately controlled and is more concentrated at the weld, preventing excessive heating of surrounding material. Lasers can be focused small enough to drill as many as 200 holes in the head of a pin. Lasers are used to trim, anneal, and microsolder the ultrasmall and very complex microcircuits that make up our modern computers, televisions, radios, and almost all types of electronic systems. In one very important research project, using the most powerful lasers ever built, scientists heat pellets of frozen deuterium (similar to hydrogen) to temperatures of millions of degrees in an attempt to initiate fusion reactions. Fusion is the process that goes on in the sun to produce its incredible energy release, and is the same process that gives the hydrogen bomb its power. If we can harness and control fusion heat to produce electricity it could provide mankind with all the energy we need.

The focused energy from low-power lasers is being used to gently and accurately repair or remove tissue from the eye and thus to correct many vision problems. The same type of system can remove brain tumors without disturbing nearby brain tissue, making the surgery safer, with less chance of the complications that can occur when surgeons use scalpels. Until recently, surgery has been done with visible and infrared lasers. The advantages of focused laser beams over scalpels are numerous. The cuts are clean and tend to bleed less, because of burning of the tissues. Now scientists are investigating the use of high-power excimer lasers with ultraviolet output, which have advantages over the other lasers and conventional surgical methods. The ultraviolet lasers can remove tissue accurately by decomposing the tissue instantly, leaving very clean cuts without burning adjacent tissues. This helps to lower the chance of unwanted damage or complications.

The coherence property of laser light has given rise to the new technical field of holography. The generation of coherent light has made it possible to record in photographic emulsion more information than is stored in a photograph. In a photograph we record the amplitude information in a light-wave front that comes from the object photographed. In holography we can record both the amplitude and the phase of the light waves coming from an object. This makes it possible to recreate the waves from the hologram just the way they originally left the object; thus when we view the re-creation of the wave front from a hologram, we see the object just as it is when we look at it; in other words, as the three-dimensional image of the object. We can move our head around and see the changes that occur as we view the object from different angles. The coherent light from a laser illuminates an object and reflects from it. The way the light is organized remains constant long enough for a photographic emulsion to record the complete wave front in both amplitude and phase, whereas with ordinary light the phase information is lost almost as soon as the light is reflected from the object. The details of how holograms are formed and images recreated from them are discussed in chapter 7.

Besides making it possible to display objects in three dimensions, another use of holography and optical interference makes it possible to record images that can reveal changes in the surface of an object to within fractions of a wavelength of light, to within millionths of an inch. This makes it possible to see strain patterns in tires, clutches, transmissions, computer disk-drive mechanisms, and precision machine parts. Seeing the strain patterns makes it possible to improve designs to make the parts perform better and last longer. Holography is being used in high-speed aircraft to display instrument readings that appear directly in front of pilots so that they do not have to redirect their vision for even a moment.

The availability of light pulses from lasers of very short duration and very high power levels has proven unexpectedly useful. This property of lasers simply is not possible to generate in any other way. Laser light pulses have made possible the attempt to initiate fusion. Research is continuing to develop lasers that can produce shorter and shorter pulses. These pulses are being used by scientists to study the behavior of atoms and molecules within very short time periods. This capability permits the study of ultrafast chemical reactions like those involved in photosynthesis and of the short-life molecules that occur in these reactions. These studies give us new information about how atoms and molecules fit together and about how they behave, so that we can, for example, create new materials that are stronger, are lighter, and last longer.

Lasers play an even more important role in communications technology. Light beams carried in cables made of glass fibers now transmit telephone conversations by the millions, signals from hundreds of television stations, and computer data to and from thousands of computers. Since these *fiber-optic* cables are expensive to install, the more information they can carry, the more worthwhile and practical they are to use. The amount of information a fiber-optic cable can carry depends on how fast the lasers used to generate the light signals can be modulated (turned on and off). Commercial lasers are available that can be modulated at 1 or 2 billion times per second. One fiber-optic cable can carry information that used to require 20,000 copper telephone wires.

Short pulses from lasers make it possible to better control the use of the laser energy. In surgery, for example, a surgeon can fire single pulses of light to repair or remove tissue and can constantly assess the results of his work before continuing. Pulses of light from lasers are being used to remove contamination and oxidation from priceless and irreplaceable works of art without damaging their surfaces.

These are just a few of the thousands of valuable uses of lasers. Lasers have made our lives richer and will continue to do so. They have provided an impetus that has accelerated the advancement of many areas of technology and science. They will make possible new and exciting developments for our future. In the chapters that follow we will discuss concepts that will help in the understanding of how lasers work. We will present in more detail the many fascinating things lasers can do. But first, perhaps you would like to know about how the laser was invented and something about the controversy that continues over who really invented it and who has the legal right to benefit financially from the uses and developments of lasers.

3

Who Invented the Laser?

SCIENTISTS knew enough about light, atoms, and molecules to build a laser as early as 1920. We knew about atoms and their surrounding electrons and the fact that the electrons in their energy levels would selectively absorb and emit light. We knew that the light could be emitted spontaneously or by stimulation from other light. We also knew about optical resonators, but at the time, they took the form of interferometers, which were used to split and recombine light beams to measure their wavelengths. However, the phenomenon of stimulated emission had not been extensively researched, and it was considered a weak effect, from which high power output was unlikely to be obtained. We knew that atoms could absorb energy, sending their electrons into excited states. The concept that remained undetected for so long was that only a correct combination of several ingredients will produce a laser. The combination requires the use of: (1) a suitable material in which excited-electron energy levels have a relatively long lifetime; (2) excitation by electrical, optical, or other means to push more electrons of atoms into an excited state than there are atoms with electrons in an unexcited state, or population inversion; and (3) an optical resonator to feed the optical field back on itself for additional amplification and building up in power. Since laser action had never been observed in nature, knowledge of the right combination of ingredients was difficult to come by. As a matter of fact, our knowledge of what occurred in nature was somewhat deceptive.

Scientists were accustomed to dealing with systems that were in thermal equilibrium. This means that materials give off light because of their temperature and that temperature determines the intensity and color of the light output. A light bulb filament operates at about 2,900°C and puts out about 550 watts per square centimeter at its surface. The surface of the sun has a temperature of 5,500°C and puts out about 6,000 watts per square centimeter at its surface. All the sources of light that had been known up to 1960 put out light with an intensity related to temperature by equations of thermal equilibrium. The laser is in anything but thermal equilibrium. Thermal equilibrium predicts that at higher electron energy levels, there are fewer numbers of atoms that will be excited to that level. Population inversion, in which there are many more atoms with electrons in excited energy levels, is far from a state of equilibrium. Nonthermal emission opens up the possibility of generating enormous light intensities without having to achieve very high temperatures. As a matter of fact, a 20-watt continuously operating laser at normal temperatures is capable of producing a light intensity of 4 kilowatts per square centimeter and when focused by lenses produces 300 kilowatts per square centimeter. A pulsed laser is capable of producing hundreds of megawatts per square centimeter and when focused produces hundreds of gigawatts per square centimeter. These are optical intensities never before observed anywhere in the universe.

So how were all the right ingredients put together at the same time? It began in several places and with several people, both in the United States and in the Soviet Union. In the United States in the early 1950s, Dr. Charles Townes was working in the field of microwaves. As we shall see in chapter 4, microwaves are high-frequency, short-wavelength radio waves, but they are much longer in wavelength than visible light. Radio waves, microwaves, infrared light, visible light, and ultraviolet light are all electromagnetic waves. Scientists usually refer to them as *electromagnetic radiation* (not to be confused with ionizing radiation from nuclear sources, which is not the same thing). They behave differently because of their wavelengths, the distances between crests of waves. In visible light the wavelength determines the color we see.

A great deal of work was being done in microwaves because of World War II and the interest in radar. Radar sends out microwaves and then looks for reflections of these waves from surrounding objects. The government was seeking sources of higher-frequency, shorter-wavelength radio waves that could better serve the purpose of radar by traveling longer distances without spreading out and losing their power. (We know this to be one of the important properties of lasers.) Dr. Townes theorized that perhaps atoms or molecules could act as electrical circuits and be used to amplify radio waves. He knew that the ammonia molecule possessed states of vibrational energy levels that might be suitable to generate radiation at the desired short radio wavelength of approximately 1.2 centimeters (about ½ inch). In 1954 Dr. Townes was able to separate excited molecules in gaseous ammonia from the unexcited molecules and use the excited molecules inside a resonant cavity to amplify and oscillate a microwave beam. The device is called a *maser*, a term that stands for *m*icrowave *a*mplification by *s*timulated *e*mission of *r*adiation. As it turned out, that particular wavelength was not good for radar because it was absorbed by water vapor in the atmosphere, and the purity of the microwave signal was not particularly well suited to communications. However, the concept paved the way for new developments, and the development of the laser was quick to follow. The right combination of ingredients had finally been put together. Molecules were being used as amplifiers.

The scientific community immediately seized on the principles that were now revealed and began looking for more dense materials that would be capable of producing more power. They were also looking for energy levels that would produce much shorter wavelengths of radiation. Their researches began to center on materials such as the solid-state substance ruby, which is an aluminum oxide crystal filled with chromium atoms. Chromium is the lasing atom; the aluminum oxide crystal simply serves as a host material. Chromium was of interest because it is a paramagnetic substance, which means that energy levels can be created by magnetic fields; these levels could be useful at microwave wavelengths. More important, scientists suspected that because of its fluorescence, ruby might be useful for light amplification. It can absorb light from ordinary sources containing wide ranges of wavelengths (broadband sources) and reemit red light. In the meantime, while the interest in ruby grew, Dr. Nicolaas Bloembergen at Harvard University had designed a continuously operating maser using a solid-state material. Soon thereafter three Bell Laboratory scientists constructed an operating device based on Bloembergen's design. This further confirmed the possibilities for amplification of an optical signal.

In December 1958, Dr. Townes, then at Columbia University, and Dr. Arthur Schawlow, at that time at Bell Laboratories, published the first paper on how a laser (then visualized as a maser operating at optical wavelengths) might be made. They had all the right ingredients but still had not been able to get a laser to operate. To everyone's surprise, Dr. Theodore Maiman, of Hughes Research Laboratories, in May 1960 built and operated the world's first laser, a ruby laser. He used a very powerful flash lamp to excite the chromium atoms in the ruby crystal. The result was an intense pulse of red light. This was the first time it had ever been possible to amplify a light beam. Mankind had been able to amplify microwaves, radio waves, and electrical signals but never before had been able to amplify a light wave. We had been able to generate light but never to add more energy to it. The atoms were behaving like billions of tiny light amplifiers.

At the same time, Drs. N. G. Basov and A. M. Prokhorov in the Soviet Union had been working with very similar concepts to construct both masers and lasers. It is not possible to know just who did what, or who did it first, or how much they were influenced by one another, but Drs. Townes, Basov, and Prokhorov were awarded the Nobel Prize in Physics in 1964 for their efforts in leading to the development of masers and lasers.

However, the story is not as straightforward as it may sound. In 1957 Gordon Gould was a graduate student at Columbia University, where Dr. Townes was a professor. He, too, worked on concepts of optical pumping to obtain population inversion and conceived of the need for an optical resonator to make the whole thing work. He realized the practical importance of what might result if lasers could be developed. He may have even been the first person to use the term laser. His notes show that he was the first person to envision that the laser might be used to start fusion reactions that could become almost unlimited sources of energy. However, he did not publish his work, instead writing it out in notebooks, which he had notarized. He applied for patents on the laser concepts but was incorrectly told that he had to build a working model in order to get a patent. An unfortunate string of events prevented him and the company he later worked for from building a laser before others did. In June 1959, Townes and Schawlow filed for a patent on laser concepts that Gould filed for nine months later. Since that time, an enormous court battle has raged over the rights to laser concepts. Gould failed to win any rights until 1977, when he was granted two patents, one on laser pumping and another on materials processing with lasers. These patent rights are still being disputed today.

Laser development proceeded at an extremely fast pace from the important moment in 1960 when the first laser operated. No doubt there were many other researchers working on lasers, and it would have been only a short period of time before other lasers would have been operational. Before the end of 1960, there were several other lasers operating in the United States. Peter Sorokin and Mirek Stevenson at IBM built and operated two lasers using rare-earth salts, uranium and samarium, in a calcium fluoride crystal as the lasing medium. These were the second and third lasers ever to operate. Soon thereafter Ali Javan, William R. Bennett, and Donald R. Herriott at Bell Laboratories developed the first gas laser using a mixture of helium and neon gasses. This type of laser can operate continuously instead of in pulses. The first helium-neon (HeNe) laser produced infrared light, which is longer in wavelength than red light and therefore not visible to the eye. It was soon discovered that this laser could also output visible red light. Because of its simplicity and reliability, the HeNe laser has become one of the most popular lasers in use today. Some models can be purchased for a few hundred dollars.

In 1962, while working for General Electric, Robert Hall

headed a group of scientists that first demonstrated that lasers could be made from semiconductors, the same materials used to make transistors and integrated circuits. This discovery opened the way for very simple and very small lasers. Research groups at IBM's Watson Research Center and MIT's Lincoln Laboratories developed semiconductor lasers at almost the same time. These first semiconductor lasers had to operate pulsed and had to be cooled to liquid nitrogen temperatures. They did not last very long. Today's modern room-temperature, continuously operating diode lasers were not perfected until 1970.

In 1963, C. K. N. Patel at Bell Laboratories, while investigating molecular systems, demonstrated laser action in carbon dioxide (CO_2). The very first CO_2 laser put out 150 milliwatts in the infrared region of the spectrum. Within two years, 200-watt CO_2 laboratory lasers had been built. In the same year, Patel observed laser action in carbon monoxide (CO), which also emits in the infrared region of the spectrum.

While working on his Ph.D. thesis at the University of Utah in 1965, William Silfvast demonstrated lasing of blue visible light in a mixture of helium and vaporized cadmium metal, the first metal-vapor laser. Many metal-vapor lasers have been invented since then and are important sources in the blue and ultraviolet regions of the spectrum.

Peter Sorokin's uranium and samarium lasers are not often used today. However, in 1966, while working with organic dyes to make laser accessories, he discovered that the dyes can be made to lase. Today the dye laser is one of the most important and frequently used lasers. It is one of the few lasers that can be made to lase at many different wavelengths.

Soon thereafter, Bill Bridges, working at Hughes Research Laboratories, developed lasers that operate with noble gas ions like argon and krypton. Ions are atoms from which one or more electrons have been removed. The same energy sources, if strong enough to excite electrons to higher energy levels, can also ionize atoms by completely removing the electrons. These continuous lasers produce output at a variety of visible wavelengths in the blue, green, and red regions of the spectrum. There are hundreds of types of gas lasers. They range from the simple helium-neon lasers that put out small fractions of a watt to huge gas-dynamic carbon dioxide lasers that can produce many kilowatts of infrared light both continuously and pulsed.

In 1975, while working for AVCO-Everett, J. J. Ewing developed the excimer laser. The term excimer refers to a whole class of lasers. The different types of excimer lasers, which are characterized by the gases used, depend on the presence of molecules of gases that do not exist unless there is a state of energy excitation. Excimer lasers may be one of the most important lasers evolving today. They are very important sources of high-power radiation in the ultraviolet region of the spectrum. We will see why they are important in chapter 7.

At Stanford University in 1975, John Madey developed the first laser that does not require atoms. It uses only electrons. Called the free-electron laser (FEL), it operates by wiggling electrons back and forth very rapidly using magnetic fields. This motion causes them to emit radiation.

As you can see, developments occurred very rapidly once the first laser was invented, and they have not slowed down much since then. The invention of the laser greatly expanded two fields of research, those of spectroscopy and holography, and gave birth to a new one, that of nonlinear optics. Spectroscopy is the study of the behavior of atoms and molecules by exciting them with light and observing how they absorb or reemit the light energy. The availability of monochromatic (single-colored), intense sources of light and the availability of these sources at many different wavelengths has thoroughly revolutionized the field of spectroscopy. Dr. Arthur Schawlow, at Stanford University, was one of the first to realize that lasers could be applied to this important field of science. His work was the beginning of the field of laser spectroscopy. Holography, the science of three-dimensional images, began in 1948 with the work of Dr. Dennis Gabor. Holography is discussed in chapter 7.

Bloembergen was one of the first to demonstrate that the very intense light from lasers could be used to do some very unusual things. Infrared light from a laser was beamed through a special crystal and, incredibly enough, turned green. The photons of the light were being converted directly from one wavelength to another. The wavelength of the infrared light was being halved. This phenomenon is called frequency doubling, since halving the wavelength doubles the frequency. Two photons of the original light are being combined to create one photon with twice the frequency and hence, as we shall see later, twice the energy. These effects involving direct changes in optical wavelength are called nonlinear effects, and the field of research is called nonlinear optics. *Nonlinear* means that the effects can only be explained by higher-order equations involving the squared or cubed value of an electromagnetic field in which the wavelengths of the fields do not remain the same. Linear optical phenomena can be described by equations that are proportional to the electromagnetic field. Soon after the initial discovery, many different types of nonlinear optical effects were found, such as frequency tripling, in which the wavelength is cut by a third; parametric up-conversion, in which photons are mixed from two different frequency sources to generate photons at the difference frequency of the two sources; and parametric oscillation, in which a photon is broken up into two photons whose frequencies are less than the original, resulting in longer wavelengths. Frequency doubling and tripling has many important uses, because each photon of the new light has much more energy than the original photons and interacts differently with atoms and molecules. For a long time there were not many blue or ultraviolet lasers, and frequency doubling and tripling were used instead to generate coherent short-wavelength radiation.

Chapter 6 will cover in detail the various types of lasers: how they are constructed, how they operate, and what types of light they emit. In chapter 7, we will discuss what can be done with laser light. In the meantime, let's talk about a few scientific principles to help us understand how lasers work.

Things We Need to Know
to Understand Lasers

Scientific Notation

To DISCUSS the concepts associated with lasers, such as wavelength, frequency, energy, and power, we must be able to indicate numbers that are sometimes either very large or small. There is a special method for doing this that helps to avoid writing many zeros after decimal points or many zeros and commas in large numbers. It is called *scientific notation.*

The number 0.1 can be written in scientific notation as 10^{-1}, which means 1 divided by 10. The number 0.01 is written as 10^{-2}, which means 1 divided by 10 twice. Thus, the number 0.00000000001 is much more easily written as 10^{-11}.

For large numbers, the notation system is similar. The number 10 is 10^1, the number 100 is 10^2, the number 1000 is 10^3, and so forth. This literally means 10 times itself (10) once, twice, three times, and so forth. Thus, the number 1 million (1,000,000) can more easily be written as 10^6.

There are also scientific names for certain commonly used scientific notations. These are used as prefixes to the units of measure. They are used for quantities like wavelength, frequency, time, energy, and power. For example, one millionth of a second is 10^{-6} second or one microsecond. The table on page 23 gives a few of these names and prefixes.

The first letter of most prefixes can be used for abbrevia-

Scientific Notation and Prefixes

Number	Scientific Notation	Scientific Name
.000000000000001	10^{-15}	femto (f)
.000000000001	10^{-12}	pico (p)
.000000001	10^{-9}	nano (n)
.000001	10^{-6}	micro (μ)
.001	10^{-3}	milli (m)
.01	10^{-2}	centi (c)
1,000	10^{3}	kilo (k)
1,000,000	10^{6}	mega (M)
1,000,000,000	10^{9}	giga (G)
1,000,000,000,000	10^{12}	tera (T)

Source: Compiled by author.

tions for units of measure. For example, millisecond is ms, kilogram is kg, and megawatt is MW. Usually, lowercase letters are used for smaller units of measure, and uppercase letters for larger units of measure, but this is not a strict rule.

Units of Measure

We will be talking about concepts such as wavelength, frequency, linewidth, pulse duration, energy, and power. It is useful to define the units of measure used for these concepts and to become familiar with some of the very large and small values that occur in dealing with lasers. The units of measure for each concept are given as they are introduced in the paragraphs that follow.

WAVELENGTH

Light normally behaves as waves, or ripples on water. The wave nature of light is shown in figure 4.1. The distance between adjacent crests or troughs, called the *wavelength,* is the fundamental parameter that describes a wave. It is the wavelength that determines how the light looks to us, that is, how we see its color. Wavelength also determines how the light behaves in physical interactions with optical materials, optical components, and detectors. Red light is longer in wavelength than blue, for example. It is interesting to note that light on a microscopic scale does not always behave like waves; sometimes it behaves as if it is made of little bundles or corpuscles, that we call photons. When one of the electrons of an atom changes states, it emits one photon of light. There are detectors sensitive enough to detect light as individual photons. However, when light travels through the vacuum of space, it behaves like a wave.

A wavelength is usually given in meters. One meter is about 39 inches, just slightly longer than a yard. A millimeter is about ⅟₃₂ of an inch. A micrometer, often called a micron, is one thousandth of a millimeter, and a nanometer is one thousandth of a micrometer. Wavelengths of visible light are expressed in nanometers. You can see that we will be talking about a very small unit of length. Twenty thousand waves of light 500 nanometers long would fit within one centimeter (about 0.4 inch).

Wavelengths of light range from fractions of nanometers to many kilometers. However, we use names other than "light" when we talk about light that is much shorter or longer in wavelength than the light we can see with our eyes. Figure 4.2 is a table of different types of "light" and their wave-

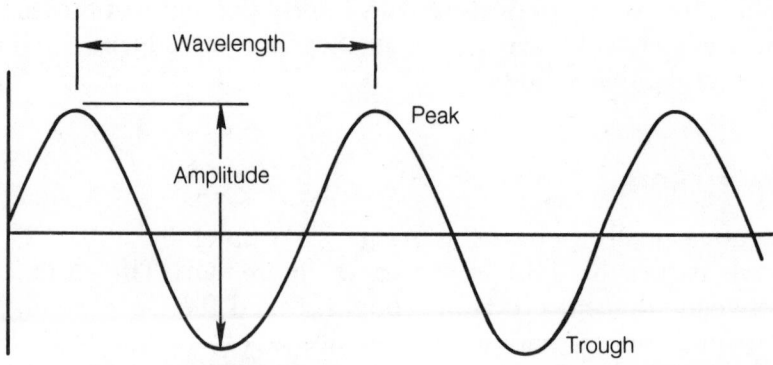

Figure 4.1. The Wave Nature of Light

lengths. They are all the same entity—light—but scientists have a more general term to cover them. They are called *electromagnetic radiation* because light is composed of rapidly oscillating electric and magnetic fields. We give different wavelengths different names because they are unique in the way they behave, the way they are used, the way they are generated, and the way they are detected. From now on we will use the term light to mean visible light, and will use the specialized names for the other types of electromagnetic radiation given in figure 4.2. Figure 4.2 illustrates wavelengths and frequencies of most of the electromagnetic spectrum and shows locations of some of the more important laser emissions.

FREQUENCY

The *frequency* of light is the number of cycles of the wave that pass a given point per second. We must understand frequency in order to talk about *bandwidth* and *linewidth*, terms used to describe how single-colored (monochromatic) laser light is. The frequency of light is simply the velocity of light divided by the wavelength. The shorter the wavelength, the higher the frequency, and vice versa.

The velocity of light is one of the most speculated upon and mysterious quantities in nature. First, it is incredibly fast. All electromagnetic radiation travels at approximately 3×10^8 meters per second (m/s). That's about 186,000 miles per second—the fastest speed in the universe. At this speed, light could travel 7.4 times around the earth in one second, to the moon (250,000 miles, or 460,000 kilometers) in 1.3 seconds, and to the sun (93,000,000 miles, or 172,000,000 kilometers) in 8.3 minutes. Einstein's theory of relativity states that no object can be accelerated to reach this speed. Electromagnetic radiation travels at this speed, and it travels only at this speed. The speed of light is perhaps one of the most universal of all constants. No matter how fast we travel toward or away from an object that sends light to us, the speed of light is always the same. Even though light is this fast, future space travel in the solar system will be hampered by delays up to many minutes for radio communications.

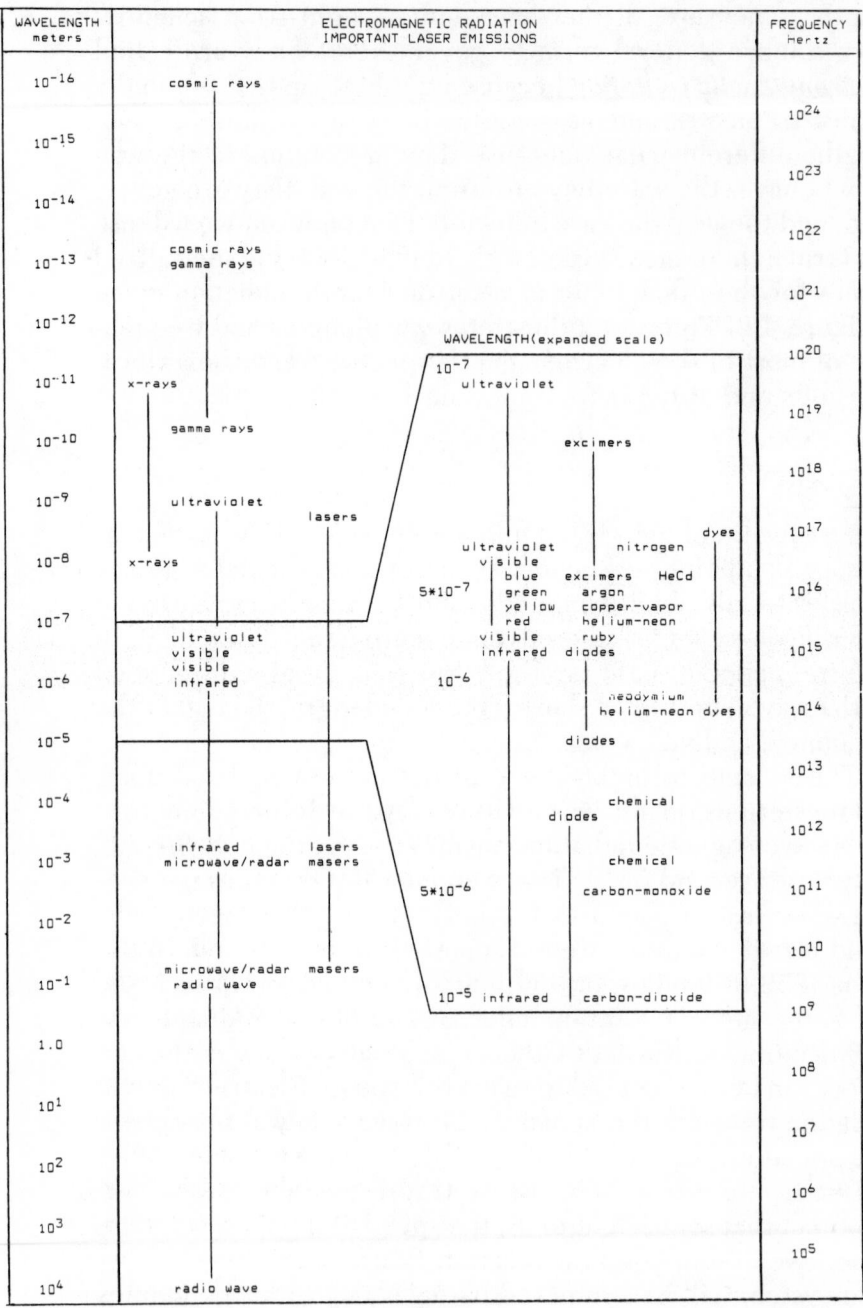

Figure 4.2. The Electromagnetic Spectrum and Important Laser Emissions

Later in this book we will talk about laser light pulses of incredibly short time duration, less than one picosecond (ps). To get some feel for how short these time durations are we can use the example of light, the fastest moving phenomenon known. Light travels 300 meters in a microsecond, 0.3 meter in a nanosecond, and only 0.3 millimeter in a picosecond.

Since the frequency of any electromagnetic radiation can be found by dividing the velocity of light by the wavelength, and since the velocity of light is a very large number while the wavelengths of visible light are very small, the frequencies are enormous numbers. Some typical frequency values are given in the following table:

Frequencies of Visible Light

Color	Wavelength	Frequency
Ultraviolet	350 nanometers	860 terahertz (10^{12}Hz)
Blue	480	620
Green	540	550
Red	700	430

Source: Compiled by author.

LINEWIDTH (BANDWIDTH)

Linewidth is the spread of frequencies (breadth of color) that is in a light signal. It corresponds to the concept of *bandwidth* used in connection with microwaves, radar, and radio waves. It arises from the fact that all sources of electromagnetic radiation are imperfect. This means that they cannot produce a perfect wave of exactly one frequency and wavelength as was shown in figure 4.1. Instead they produce a wave that is more correctly shown in figure 4.3. They produce signals that are not exactly one wavelength (monochromatic) but that are a mix of signals with slightly differing wavelengths. In the areas of microwave, radar, and radio, bandwidth is very important, and we purposely create signals with wide bandwidth. The bandwidth determines how much information can be carried by signals, for example, by phone calls and computer data. When laser light signals are used to carry information, linewidth (bandwidth) becomes important. However, for now, we need

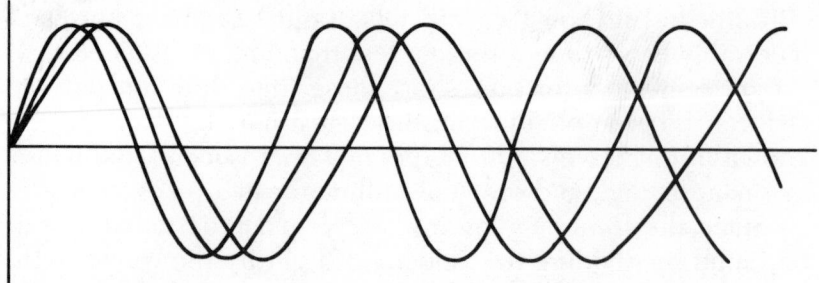

Figure 4.3. Multiple Wavelengths (Frequencies) in a Light Signal

to understand that lasers have unusually narrow linewidths in comparison to ordinary sources of light. This is one of the special features of laser light that makes lasers so important to research and industry.

Linewidth, as we have said, is a measure of the laser's single-coloredness, or the monochromaticity of a light signal. Let us look at some examples for visible light.

The light we perceive to be green runs from about 510 to 580 nanometers. This corresponds to frequencies of 588 to 517 terahertz. The difference is 71 terahertz or about 10^{14} hertz. Before lasers were invented, scientists obtained monochromatic light using optical devices like diffraction gratings or prisms to spread the many wavelengths that come from ordinary sources of light over a wide area, selecting the desired wavelength by the use of a thin slit. The linewidth possible with this method is approximately one gigahertz. This is about 1/10,000 of the total linewidth of green light. Once the light is spread out and a part of it picked off with a slit, very little power is left. The more monochromatic the light, the narrower the slit and the weaker the power that can get through. Let's compare this to a laser.

One of the most monochromatic sources yet devised is a highly stabilized and continuously operating dye laser with a narrowed linewidth of 100 hertz, a 10-millionfold improvement over using a grating. At an operating wavelength of 298 nanometers, this laser corresponds to a wavelength spread of 3×10^{-11} nanometers. It can also deliver 1 watt of power at this linewidth, roughly a million times more power than with the grating.

This new kind of light has properties radically different from anything known before. We will see in chapter 7 the incredible new possibilities opened up by these properties for scientific investigation and practical applications.

TIME DURATION

An important characteristic of lasers is the fact that they can emit their light in very short pulses. They can emit single pulses, one pulse of light each time they are fired; they can emit trains of pulses, one short pulse after another; and they can operate continuously. The terms *pulsed* and *continuous wave* are used to describe these conditions.

In single-pulse operation, the lasing medium is flash-pumped very quickly, establishing a strong population inversion. Then the resonator is quickly brought into a resonant condition (we will discuss later how this is done) to help channel all the energy of the population inversion into a single short pulse. With this method, very short pulses at extremely high power levels can be produced. One of the most important applications for single pulses is for fusion initiation, to be discussed in chapter 7. Extremely short-duration pulses are very important in scientific research.

Lasers now routinely emit trains of pulses with durations of around one picosecond. More recently, pulses have been generated that are only fractions of a picosecond (in the femtosecond range).

In continuous-wave operation, lasers emit light continuously, as a light bulb does. The output power is nearly constant. Typical commercial lasers have output stabilities of 1 or 2 percent.

ENERGY

Energy is a measure of the content of electromagnetic radiation. The energy of a wave is proportional to the amplitude of the electromagnetic field that forms the light. On a microscopic scale, the energy in a light signal is the number of photons in a light beam passing a given point in a given length of time. Energy is measured in units called *joules,* which can be abbreviated as *j.* A 100-watt light bulb puts out 100 joules of energy each second. Some of its power is in the form of heat; what we experience as heat from the bulb is actually infrared radiation that our bodies absorb, converting it to heat. An object weighing ½ kilogram (about 1 pound), dropped a distance of 0.2 meter (about 8 inches), has about 1 joule of kinetic energy before it hits. A 3,500-pound automobile moving at 30 miles per hour has about 143 kilojoules of kinetic energy.

How much energy is in a single photon? The energy in a photon is given by a constant, called Planck's constant (abbreviated as h), multiplied by the frequency of the light. Planck's constant is 6.6×10^{-34} joule-seconds. Thus, the energy in 1 photon of green light is h times 600×10^{12} or about 4×10^{-19} joules. Since the energy in a photon is proportional to frequency, the photons with the shortest wavelength are the most energetic. The following table gives some typical photon energies for various types of electromagnetic radiation.

Single Photon Energies

Type	Wavelength		Energy
Gamma rays	5	pm	40.0 femtojoules (fj)
X-rays	0.5	nm	.40
Ultraviolet	100	nm	.002
Visible	500	nm	.0004
Infrared	5	μ	.00004

Source: Compiled by author.

An X-ray photon has 1,000 times as much energy as a green photon. And what do we use X rays for? To penetrate through material, to look deeper inside. In research, with shorter wavelengths, more energetic photons are being used to penetrate deeper into atoms and molecules to discover how they behave. Recently-developed high-power ultraviolet lasers, called excimer lasers, have taken over many applications that were being done by visible and infrared lasers. The UV (an abbreviation for ultraviolet) laser cuts certain materials more quickly, more cleanly, and without burning. This is important, for example, in medicine for removal of tissue. A great deal of research is now in progress to develop even shorter-wavelength lasers.

A 1-watt, 500 nanometer-wavelength laser beam that produces 1 joule per second sends out about 2.5×10^{18} photons per second (2,500 teraphotons/sec). Thus a photon of light is a very small bundle of energy. In everyday life we would never have to be concerned about the fact that light is composed of photons, but in research it is very important. For example, astronomers attach to telescopes instruments that are sensitive enough to look at the light on a photon-by-photon basis.

POWER

Power is the rate at which energy is produced. The unit of power is the watt. A watt is the number of joules produced per second. A 100-watt light bulb produces 100 joules per second, containing about 10 watts of visible light and about 90 watts of heat (infrared light). A laser also puts out light and heat, but the heat is mostly confined to the lasing medium and is not emitted with the laser beam. When we speak of a 1-watt laser, we are talking only of the power of the light output; the total power required to excite the medium may be as much as 1,000 watts for a 1-watt laser. Thus, a 1-watt laser appears to be very bright compared to a 1-watt light bulb. A note of caution. It may be safe to see laser light when the light is scattered from objects other than mirrors in a room. But never look directly into the laser, the laser beam, or a reflected laser beam. If the eye focuses even a small fraction of a watt of laser energy on the retina, it will burn and damage the retina, inducing permanent, partial, or full loss of sight. Infrared lasers require particular care since their output is not visible.

When we talk about the power output of lasers, the numbers we will deal with will be very different depending on whether we talk about continuous output or the power output in a single pulse. A continuous-wave laser at 1 watt puts out 1 joule per second, but a pulsed laser producing 1 joule of energy in a nanosecond has a pulse power of 1 joule divided by 10^{-9} second, which is 1 gigawatt. Only lasers can provide short-duration high-power pulses of this magnitude, which is another important feature that sets lasers apart from other light devices and makes them valuable in research, industry, and medicine. In the case of lasers that repeatedly generate pulses, we will talk about the laser's average power. The average power of a pulsed laser is the energy per pulse divided by the time between pulses.

Coherence

Coherence refers to the relative organization of the light waves in the light signal. The term can apply to any form of electromagnetic radiation. The laser was discovered partly as a result of the search for more coherence in microwave oscillators. Prior to the invention of lasers, all sources of light were incoherent. We can run electricity through a filament as in a light bulb or through a gas as in a fluorescent lamp, but we do not have control over how the electrons in the atoms emit their light energy. The result is a broadband emission of enormous linewidth consisting of many frequencies or wavelengths and without any particular organization in space or time. However, in a laser we can take advantage of the way certain atoms behave when their electrons are excited into quasi-stable energy states. The electrons sit there just long enough for photons of the right wavelength to come along and knock them out of the energy state. This happens when another electron in the same state in another atom falls out of its energy level and spontaneously emits a photon—the phenomenon is called spontaneous emission. The new photon will be very close to the same wavelength, will proceed in the same direction, and its wave will be in phase with the incoming photon wave. Thus, as all the electrons in the lasing medium emit photons, the resulting light signal or beam becomes highly organized.

There are two types of coherence in a light beam and they are independent of each other. To understand coherence it is necessary to introduce the concepts of *wave fronts* and *phase differences* between waves. A wave front is the surface pattern of light emitted from a laser or reflected from an object. The surfaces are perpendicular to the direction the waves are moving. Wave fronts are discussed in more detail in connection with holography in chapter 7. Phase is a measure of the difference in the position of the peaks or troughs of two waves. Phase is measured in degrees, similar to the way an angle is measured. The distance between peaks is 360 degrees. The beam can be either spatially or temporally coherent or some combination of the two. Spatial coherence is a measure of the relative phase of the waves emanating within the beam at different locations along the beam wave fronts. Temporal coherence is a measure of the constancy of phase in time of the laser beam measured at a given point. Examples of spatial coherence and incoherence are shown in figure 4.4. Notice that the coherent waves are lined up with each other. They are in phase while the incoherent waves may take on any arbitrary phase relationship.

Spatial coherence exists between the waves emanating from two points on a wave front when the phase at one point is predictable from the phase at the other. If the phase between the two points is random, then they have no spatial coherence. If the waves from the two points stay in phase as time progresses, then the two points have perfect spatial coherence. If this is true for any two points in a wave front, then the wave front has perfect spatial coherence. Generally, wave

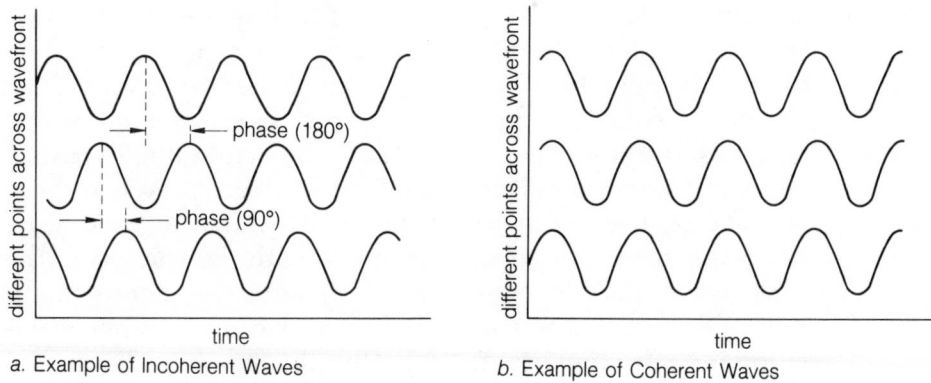

a. Example of Incoherent Waves   b. Example of Coherent Waves

Figure 4.4. Spatial Coherence and Incoherence

fronts are only partially coherent. This means there is coherence in only a limited region surrounding a given point.

Temporal coherence exists in a wave if the phase at a given time is predictable from the phase at another time. If the phase is random between the two points in time, then there is no temporal coherence in that time period. If at any time, and over a given time period, the phase difference of the field value at a given point in the wave remains the same, then there is coherence in the wave over that time period. If this is true for any time period, then the wave has perfect temporal coherence. If the time period over which this occurs is limited in duration, then the wave has partial temporal coherence. The time period over which the wave has temporal coherence is called the *coherence time*. The coherence time multiplied by the speed of light is the *coherence length*. Notice that temporal coherence is closely related to monochromaticity, since the two waves of different wavelength have constantly varying phases.

Interference

One of the ways that light behaves like waves is in the production of *interference* patterns. An interference pattern is a series of light and dark bands (also called fringes) that come about when the light waves from two different beams of light are mixed. Parts *a* and *b* of figure 4.5 show two light waves that are in phase. Part *c* shows the result of mixing the two waves. They combine to form the sum of the two waves. When the crests coincide, they form a crest twice as high as the originals, resulting in a wave of twice the amplitude. In parts *d* and *e* the waves differ in phase by 180 degrees. When these waves mix, they cancel each other. To put it another way, when the crests of one wave coincide with the troughs of the other wave, the outcome is a null, or no wave, a zero amplitude. The fact that mixing of light waves can result in cancellation is a very important and useful phenomenon. These two examples illustrate that the intensity of a mixed wave is highly dependent on the phase of the waves that are mixed.

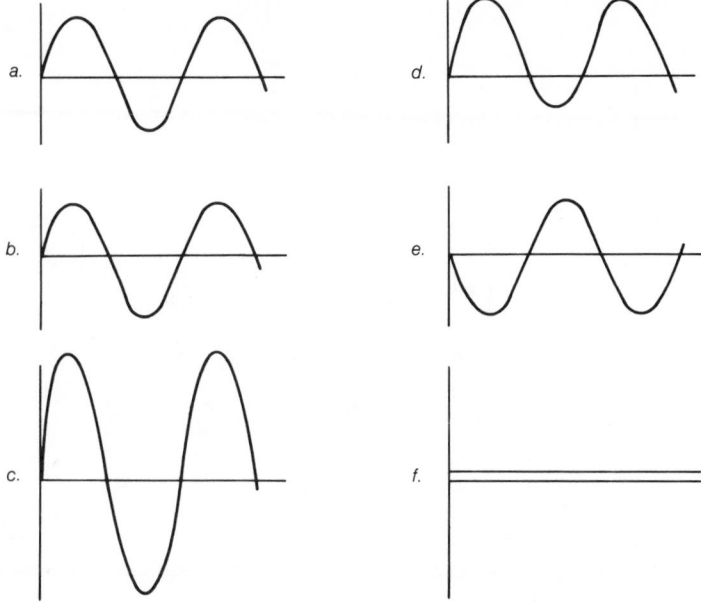

Figure 4.5. Mixing of Light Waves

A method that takes advantage of wave mixing and that generates interference patterns is shown in figure 4.6. The beam from a laser is divided by a partially reflecting mirror, M1, that redirects part of the beam to mirror M2. Both beams are projected by lenses onto a surface at location P. The two beams mix with each other at the surface. The resulting light intensity at any point along P is dependent on the phase of the two beams. This is determined by the path difference represented by 1 and 2 to a point on P. The result is an intensity along P that is shown in part *b* of the figure. The interference pattern looks like the stripes of light and dark shown in part *c*. The light regions are where the beams have reinforced each other, and the dark regions are where the beams have canceled each other. If P is not flat, the pattern changes. We can observe in the pattern minute changes in the shape of P, changes smaller than fractions of a wavelength of light. When we talk about holography in chapter 7, we will see how useful inter-

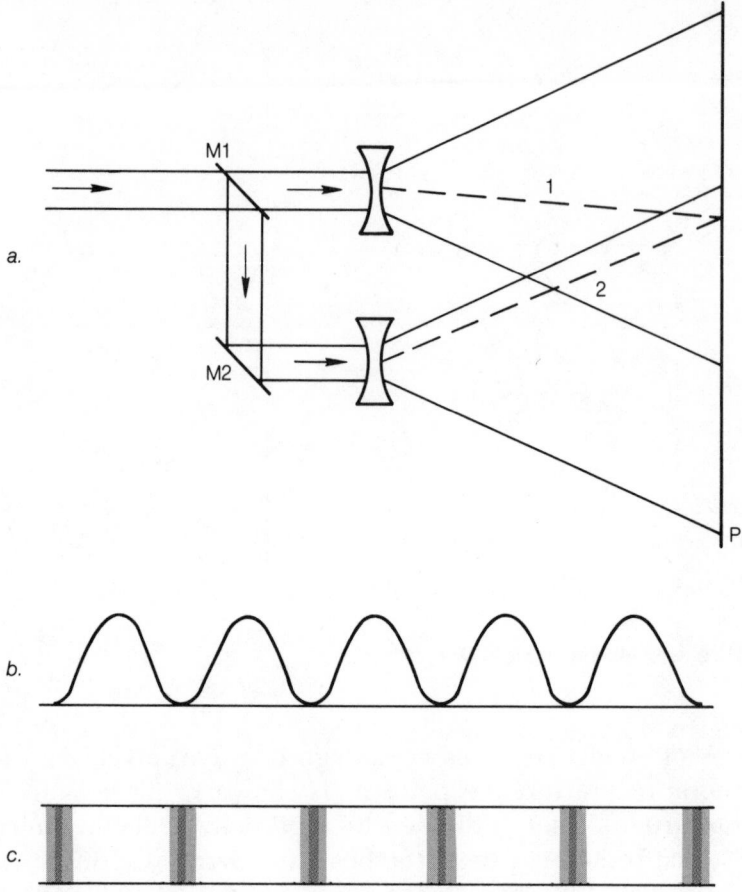

Figure 4.6. Formation of an Interference Pattern

ference is in examining manufactured parts for defects and stress weaknesses.

A simplified diagram of a very useful scientific optical device called an *interferometer* is shown in figure 4.7. An incoming light beam from the lower left is split by a partially reflecting mirror, M1, into two beams. The two beams travel along different paths, a and b, until they arrive at another partially reflecting mirror, M4, on the upper right, where they

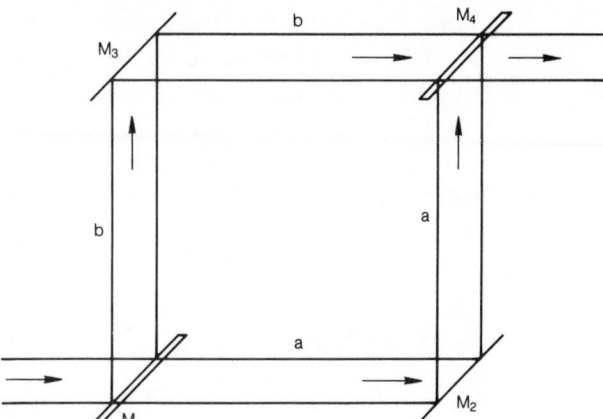

Figure 4.7. Diagram of a
Simple Interferometer

are mixed again. The result is an interference pattern at the
output of the interferometer. The position of the interference
pattern depends critically on any difference in the lengths of
paths a and b. If anything is done to change the path lengths
by an amount as small as a fraction of a wavelength, the
interference pattern will noticeably move. Thus, the interferometer offers the ability to measure distances to within
fractions of a wavelength. Since the wavelength of visible light
is about 500 nanometers, this means that we can measure
changes in path length to within less than a nanometer.

Two things should be pointed out about interferometers.
The interference pattern will not show up unless a constant
phase relationship is maintained between the two beams. An
example is the waves represented in figure 4.5. If there are
many waves adding together, and especially if they have different wavelengths, the interference pattern may not show up.
Since the peaks are lower and the cancellation is lessened, the
pattern is smoothed out. This is the reason for using only one
source of light, splitting the incoming beam into two pieces,
and remixing them. If beams from two different sources are
mixed, there is no fixed phase relationship between the two
waves, the waves are constantly changing with respect to each
other, and the interference pattern does not appear. It is not
necessary for the incoming beam of light to be from a laser.

Under the right conditions, any source of light can produce an interference pattern. However, since laser light is both coherent and monochromatic, it makes it much easier to take advantage of interference under a greater variety of conditions and to make precise measurements and observations. Many of the laser applications discussed in chapter 7 are based on interference of light waves.

Index of Refraction

When light travels in space or in a vacuum, it travels in straight lines at constant velocity. The velocity of light is usually represented by the letter c. When light travels within or between optical materials or crystals, it can change direction and velocity. The changes in direction and velocity are determined by the *index of refraction* of the optical material. The index of refraction is related to the density of the material. The higher the density, the higher the index of refraction.

The reflection and refraction of a laser beam by optical materials is illustrated in figure 4.8. The incoming or incident beam, A, is in a material with an index of refraction n_1 and at an angle a. The beam can be reflected back in medium n_1 as shown by B or can be refracted into medium n_2 as shown by C. The angle b and amount of beam reflected or refracted depends on the indices of refraction n_1 and n_2. Any text on optics can explain how the angles relate to each other and how the amount reflected or refracted depends on the indices of refraction. For our purposes, there are two important concepts concerning indices of refraction.

The first important concept is as follows. When n_1 is greater than n_2 and angle a is large enough, the entire beam reflects into B and none goes into medium n_2 to form C. This is called *total internal reflection*. It is the basis for making fiber-optic cables that can carry laser beams encoded with information over hundreds of miles. These cables are now being used to replace conventional telephone cables because the fiber-optic cables are smaller and carry more information. We will discuss fiber-optics in chapter 7.

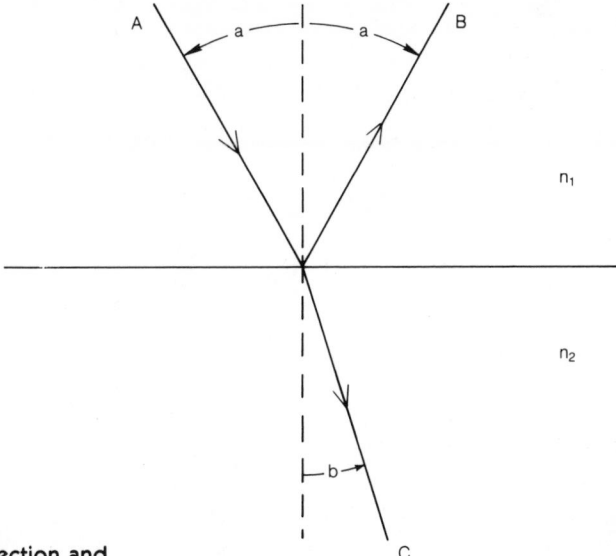

Figure 4.8. Reflection and
Refraction by Optical Materials

The second important fact about indices of refraction is that the velocity of light is different in an optical material than it is in space. The velocity of light in a material of index of refraction, n, is c divided by n (c/n). As we will see in chapter 5, this fact forms the basis for our ability to modulate (encode information in) laser beams.

Polarization

Light is made up of rapidly oscillating electric and magnetic fields. *Polarization* refers to the direction of the electromagnetic fields in a light beam. The direction of polarization is perpendicular to the direction the light is traveling, as shown in figure 4.9–a. As the light travels in direction C, the electric field in the wave can be aligned along A, or along B, or along a combination of the two. If all the light is aligned along A, it is said to be *linearly polarized* in the direction A. If all the light is aligned along B, it is linearly polarized in the direction B. Light emitted from an ordinary light bulb is a random combi-

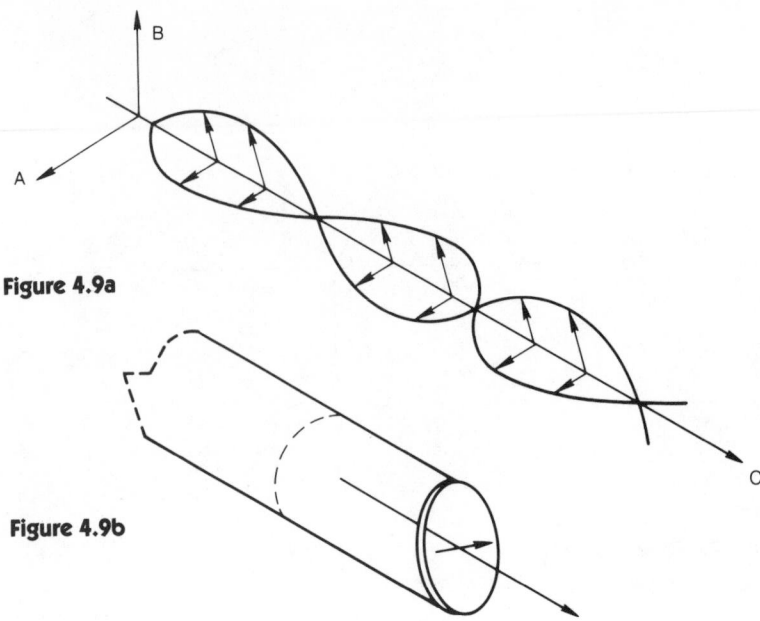

Figure 4.9a

Figure 4.9b

Polarization of a Light Beam

nation of polarizations in all directions. It is said to be *unpolarized*.

Light can become polarized in many ways. When light is reflected from a surface, as shown in figure 4.8, and provided it is randomly polarized, more of that part of the light polarized parallel to the surface of reflection (the surface between n_1 and n_2) is reflected. Further, more of that part of the light polarized at an angle to the surface is refracted into the surface. Thus, light reflected from a crystal, mirror, or glass tends to be linearly polarized. There are crystals that transmit one polarization direction but not the other directions. When randomly polarized light passes through these crystals, it becomes linearly polarized. These crystals are called *polarizers.* Some sunglasses are made of a polarizer that reduces the amount of light transmitted. The polarizer in the glasses is set to transmit vertically polarized light. Sunlight reflected from the hood of your car tends to be polarized in the horizontal direction. Thus the polarizing sunglasses tend to eliminate the reflected light more efficiently than the rest of the light coming through the windshield.

Gas lasers often have windows at each end of the gas tube to seal the tube and to allow the laser light to go between the gas and the resonator mirrors. These windows are slanted at an angle for best transmission of the laser light. The best transmission angle is called the *Brewster angle,* and the windows are called *Brewster windows.* When lasers use Brewster windows, their output light will be polarized in a direction parallel to the window's surface. The direction of polarization is shown in figure 4.9–*b.* Some gas lasers do not use Brewster windows. Instead, they use the resonator mirrors to seal the gas tube. This makes the laser simpler, and the resonator has lower losses. However, it has the disadvantage that the windows cannot be realigned or replaced.

Birefringence and Modulation

Certain types of crystals, called *birefringent crystals* because of the fixed arrangement of the atoms in their crystalline lattice, will transmit laser beams with different indices of refraction based on the direction of polarization. This means that the part of the beam polarized in direction A of figure 4.9 will move through the crystal at a different velocity than the part of the beam that is polarized along B. Therefore, if the waves in the beam were in phase before entering the crystal, they will be out of phase by a fixed amount by the time they emerge from the crystal. The amount they are out of phase depends on the length of the crystal, the values of the indices of refraction, and the wavelength of the light. Last, and perhaps most important: depending on the type of crystal, the indices of refraction can be changed by the application of acoustic vibrations (sound waves)or the application of electric or magnetic fields to the crystal. The acoustic, electric, or magnetic fields can be varied in accordance with a signal, such as a telephone conversation, television signal, or computer-data channel, and can be encoded into the laser beam. This is called *modulation* of the laser beam. As the indices of refraction vary, the phase of the parts of the laser beam traveling in the two different polarizations in the crystal are varied. The modulation of the laser beam can be set up so that the laser light varies in phase, polarization, or amplitude in response to the modulation signal. Detectors that receive the laser beam can be set up to recover the modulation signal from the laser beam. Since the light wave itself is of a very high frequency, it is possible to modulate it at much higher frequencies than can be used with wires, microwaves, or radio waves. Thus the light wave can carry much more information. Thousands of telephone conversations can be combined by electronic techniques called *multiplexing* and be carried by a single laser beam. The laser beam can be sent through the air or can be piped down a fiber-optic cable.

Ground-to-space communications may soon be possible by sending laser beams through the atmosphere into space. Communications between cities are carried best by fiber-optic cables, since obstructions and weather conditions can interfere with atmospheric transmission. The uses of lasers for communications will be covered in more detail in chapter 7.

This chapter has presented some of the fundamental concepts in optics necessary for you to begin to understand lasers. In the next chapter we will talk about concepts directly associated with laser operation. If you want more detail than we have presented here, any good undergraduate text on optics will be helpful.

How Do Lasers Work?

BEFORE we can understand how lasers work, there are a few more concepts that we must review. These are:

- Electron energy levels
- Gain
- Optical resonators
- Modes of oscillation
- Q-switching
- Mode locking
- Modulation.

Electron Energy Levels

The atoms of all materials are made up of a small, dense nucleus of protons and neutrons surrounded by a halo of electrons, one for each proton in the nucleus. Molecules are combinations of atoms held together by the bonding forces of the atom's electrons. The electrons can be thought of as traveling in orbits around the nucleus. The electrons are really not little objects moving in circles but are distributed around the nucleus, spread out in various diffuse patterns. In atoms with many electrons, the electrons fit systematically into sets of energy levels that are of specific, fixed, and discrete value with gaps between them. A given electron does not possess an arbitrary energy value but must reside at one of the specific levels. These levels arise in nature and apply to all materials because of the properties of the forces between electrons and protons, the forces between electrons and other electrons, and the wave nature of the electron.

The existence of levels in sets of discrete values is called the quantized model of the atom. It is a product of modern physics and can be used to describe the behavior of atoms and electrons, and the bonding of atoms to form molecules. For a knowledge of lasers, it is important to understand that there is a fixed, predetermined set of discrete energy levels for electrons surrounding atoms. Electrons can move from lower levels to higher levels by absorbing energy, a process called *absorption*, and they can move from higher levels to lower levels by releasing energy through a process called *emission*. The movements among energy levels are referred to as *transitions*. Some transitions involve the absorption or release of photons; they are *radiative*. Other transitions do not radiate but involve other forms of energy absorption or emission: They are *nonradiative*. When an atom is in the ground state, it is not excited by outside energy, and all of its electrons are in the lowest energy levels allowed by nature.

The electron energy levels are grouped into shells, and each shell can contain only a limited number of electrons. The shells do not have to be full. The more electrons an atom has, the more its shells are occupied. The more shells that are occupied, the higher the energy levels of the ground state. Since two electrons, and only two, can occupy the first shell of an atom, the third must go to the next higher shell. Since this shell will hold only eight, the eleventh electron must go to a higher shell, which holds up to eighteen electrons. This type of sequence continues for all atomic elements. The picture presented here is highly simplified. A complete description of electron energy levels is a very complex subject but is not necessary for an understanding of laser operation.

We will also be looking at energy levels in molecules that can act as laser mediums. These energy levels are also discrete, but there are many, and they are very close together, making them look like bands of continuous energy levels. These bands are the result of vibrational and rotational energy that can be

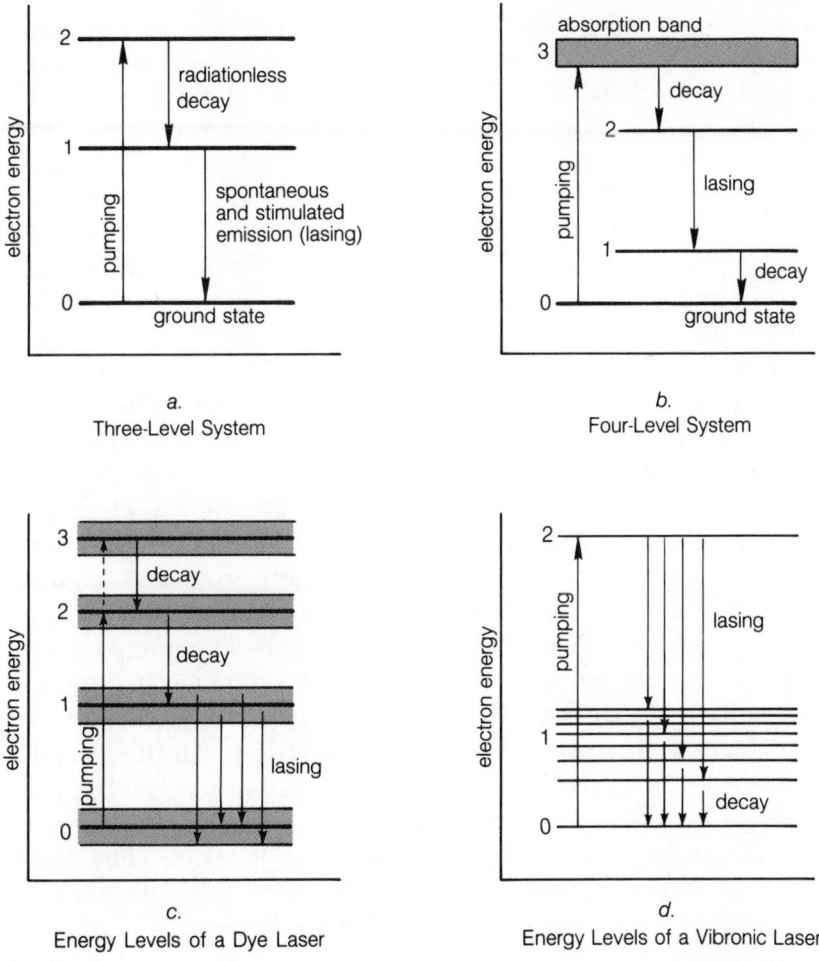

Figure 5.1. Electron Energy Levels for Various Types of Laser Media

present in molecules. Various types of energy-level arrangements are shown in figure 5.1. Here we follow the convention that defines the lowest or ground state as zero.

There are two important properties of energy levels. The first property, its population, is the number of atoms in the laser material whose electrons are on the energy level. The second important property, when an electron is excited above ground level, is the average lifetime of the energy level, which is a measure of its stability. The lifetime is the average length of time an electron boosted to a higher level will stay there. Usually the levels are unstable, and excited electrons last only about 10 nanoseconds in an energy level, but there are some levels on which they can last considerably longer. The three energy levels shown in figure 5.1–*a* provide a look at how energy levels work. All atoms have many more levels than this (there can be thousands), but we show only three for simplicity. In the ground state, all three levels are populated by electrons, and there will always be fewer electrons in level 2 than in 1, and fewer in 1 than in 0. Level 1 is always populated, but it is necessary that the population of 1 exceed 0 for lasing to occur; otherwise, the atoms would be more likely to absorb radiation than to emit it.

When the atoms are subject to excitation, such as electrons from a power source or photons from a flash lamp, the electrons can absorb energy moving from their ground states into excited states. Electrons excited by the pump source may move into level 2 or into other higher levels. The pump source may be very broadband. The electrons then may transition by nonradiative processes from level 2 and other higher levels into level 1. If level 1 has an unusually long lifetime, many atoms will suddenly have many more electrons in level 1 than in level 0. When this occurs, the gain of the medium rises as the electrons in level 1 begin falling to level 0. They emit photons whose frequency is proportional to the difference in the energy level. This process may be written as

$$f = (E_1 - E_0)/h$$

E is the energy level, f the photon frequency, and h is Planck's constant.

Thus electrons absorbing broadband energy move into upper levels, then fall through various types of nonradiative transitions into level 1. There—spontaneously and by stimulation of other photons—they can emit light at nearly one frequency in an organized manner. In short, they can lase.

The energy levels of a four-level system are shown in figure 5.1–*b*. The pump sends electrons by absorption into a band of energy levels. These levels decay by nonradiative transitions to level 2. The accumulation in level 2 becomes much greater than in level 1, and stimulated emission occurs with transitions to level 1. Level 1 is not the ground level but will also decay to level 0. This energy-level structure is typical of a common solid-state laser called the Nd:YAG, which we will discuss in chapter 6.

The energy-level diagrams of parts *c* and *d* of figure 5.1 are included to illustrate that some types of lasers are tunable, that is, they can be made to lase at various selected frequencies. The frequencies are selected by favoring oscillation at the desired wavelength by making the cavity resonant (low loss)at only that particular wavelength. This can be done by special reflection coatings or by placing a dispersive element such as a prism or grating in the optical cavity. A cavity is low loss when the mirrors are highly reflective at the desired wavelength.

The energy-level diagram of figure 5.1–*c* is typical of an organic molecule called a dye. Here the energy levels become bands as a result of many closely spaced vibrational and rotational energy levels present in molecules. Thus we have a variety of levels that can transition to another variety of levels, making it possible to lase over a band of different wavelengths. The pump excitation sends electrons into levels 2 and 3, which decay to level 1, making level 1 population inverted with respect to level 0.

Another type of tunable laser is characterized by the energy levels shown in figure 5.1–*d*. This is called a vibronic laser, because the lower lasing level is a band of energies generated by the vibrations of an atom placed in a crystal structure. This is typical of chromium in alexandrite, which we will discuss in chapter 6. The pump sends electrons into level 2, which can lase by transitions into many levels of energy at level 1. Level 1 then decays by nonradiative transitions into the ground state.

Gain

The measure of amplification in a laser medium is called *gain*. Gain is a measure of the ability of a medium to increase the intensity of a light signal as the light traverses a given distance in the medium. It is usually given as the amplification per centimeter. For example, if a 1-joule pulse has 10 joules of energy after passing through 1 centimeter of laser medium, the gain is 10 per centimeter. This can be written as 10 cm^{-1}.

An important consideration for lasers is the gain available for different wavelengths. You would think from our description that a laser operating at one wavelength has gain at only that precise wavelength, since ideally all the electrons in all the emitting atoms produce photons of the same wavelength (if they all transition from the same upper level to the same lower level). This is an idealization and there are many factors in the real way atoms and light actually behave that change this picture. A typical gain curve for a laser is shown in figure 5.2, which shows the gain available as a function of wavelength in the region of a lasing transition. The ideal gain curve would just be a line centered at the laser's operating frequency. The real gain curve has a bell-shaped profile.

There are several mechanisms that contribute to this linewidth broadening. There is a natural line breadth. This is due to the fact that the spontaneous emission process itself is imperfect. Since there is a finite lifetime of the excited state, the energy level is not precisely determined. Therefore, the atoms can emit at slightly different frequencies. This effect is small for most lasers. In gas lasers at low pressures the linewidth is primarily due to the fact that the atoms are engaged in random motion, and the frequency of the emitted radiation

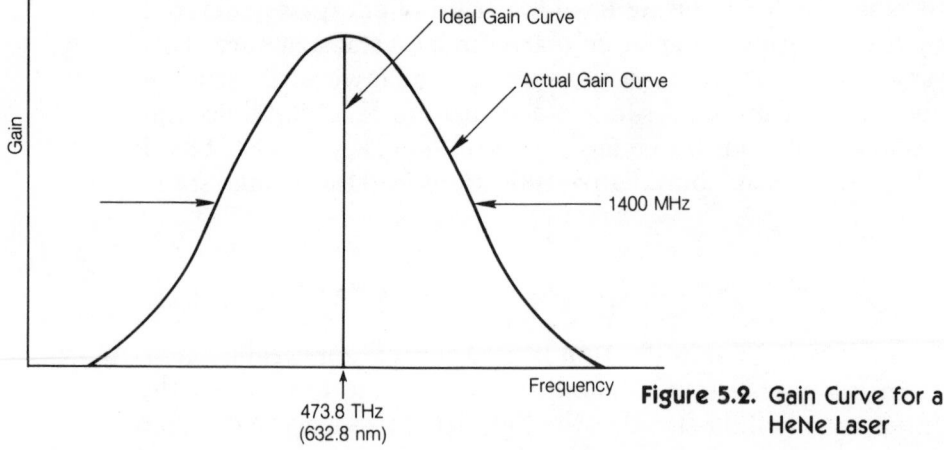

Figure 5.2. Gain Curve for a HeNe Laser

varies depending on the velocity and direction of motion. This effect is called *Doppler broadening.* The Doppler effect is the change in frequency of emitted light when a source is moving toward or away from an observer. Astronomers have been using this effect for many years to measure the speed that stars and galaxies move toward or away from us. Their light is shifted toward the blue part of the spectrum when they move toward us and toward the red part of the spectrum when they move away from us. Another broadening effect is called *collision broadening* and is prevalent in higher-pressure lasers, where more collisions take place among the atoms. These collisions interrupt the initial and final levels of the transitioning electrons.

Optical Resonators (Cavities)

The optical resonator of a laser is a key ingredient in making a laser work. It makes the laser into an oscillator by acting as a feedback circuit for optical energy. It minimizes the amount of amplification (gain) required in the laser medium to build up the optical field to an appreciable level to achieve oscillation. Once the optical field builds up, it can be sustained by stimulated emission as long as the excitation maintains the population inversion. Sometimes lasers are referred to as laser oscillators. This simply means that the laser has an optical resonator; the name distinguishes it from a laser amplifier. A laser amplifier has a lasing medium that is pumped to population inversion to amplify an incoming light signal that has been formed by a laser oscillator. The light field makes only one pass through the amplifier.

Some common laser resonators are shown in figure 5.3. The mirrors are placed with their inner surfaces parallel and reflective surfaces facing inward. The reflecting surfaces may be plane (flat) but are usually spherical so that the laser light is focused slightly into the lasing medium. The mirrors are coated for high reflectivity at several wavelengths or a broad band of wavelengths, to provide feedback at wavelengths where the medium is expected to lase. For oscillation to occur, the total

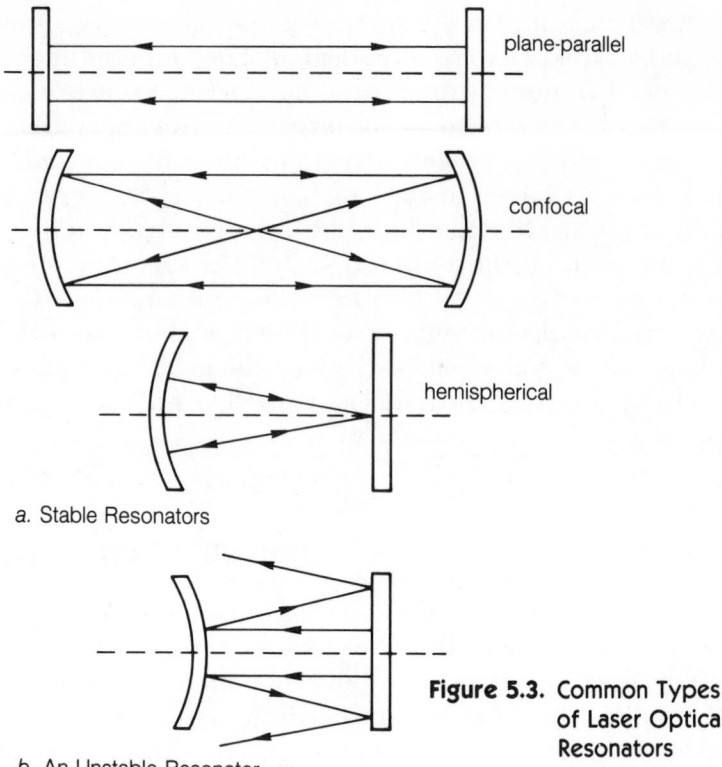

plane-parallel

confocal

hemispherical

a. Stable Resonators

b. An Unstable Resonator

Figure 5.3. Common Types of Laser Optical Resonators

gain along the length of the medium must equal or exceed the losses in the optical resonator. Some laser mediums will operate at only one wavelength and others at several wavelengths, while still others will operate over a broad band of wavelengths. Usually one mirror, the back mirror, will be coated for maximum reflectivity at the operating wavelengths. The other mirror, the output mirror, will be coated for partial reflection to permit transmission of the beam for use outside the laser resonator. The coatings, mirror radii, separations, and mirror alignments determine the losses of the optical resonator. The losses are chosen to match the gain characteristics of the lasing medium.

The losses of a resonator determine the quality of the resonator. Scientists use the letter Q to represent resonator quality. The lower the losses, the higher the Q, and vice versa.

Usually it is the Q of a resonator at a given wavelength that determines how and at what wavelength the laser operates. If the Q is high enough, the laser fires; if not, the laser does not emit light. The Q is determined by the reflectivity at the mirrors (more specifically, at the mirror coatings); by reflections from any devices or objects inside the resonator, such as a modulator or an etalon (see the section on single-mode operation for a description of an etalon); by reflections from Brewster windows; by scattering due to imperfections in the laser medium; by scattering from dust on the optical components; and finally by the alignment of the resonator mirrors. The alignment must be very precise. The mirrors must reflect the optical beam back on itself, so that in a round trip through the resonator, the optical beam reproduces itself. Otherwise, on each trip part of the optical beam is lost. Thus, reflectivity must be high at the mirrors and low everywhere else in the cavity. Because scattering must be kept low everywhere inside the resonator, all surfaces must be very clean and of high quality. And the mirrors must be kept precisely aligned. For research lasers that require very high monochromaticity, operation must be kept stable on one axial mode (see the next section), which requires keeping the mirrors exactly the same distance apart and regulating the temperature of the single-mode etalon. The mirrors are kept apart by spacing them with rods of materials such as cervitt glass, which expands or contracts very little with temperature change.

The plane-parallel resonator forms a beam sweeping out a larger region of the lasing medium. The beam is difficult to keep aligned and has losses as a result of beam walkoff to the sides of the resonator. The confocal resonator is the most common form for lasers, especially for continuous gas lasers, which are low in gain. It is fairly easy to align, is stable, and has low loss. Some special lasers, like high-power lasers possessing exceptionally high gain (see chapter 7), need to output their energy quickly, after a few passes in the resonator. For this purpose, "unstable" optical resonators are best. A diagram of an unstable optical resonator is shown in figure 5.3–b. The term *unstable* refers to the fact they have high losses, that is, they quickly give up the laser energy to the outside.

Modes of Oscillation

The use of an optical resonator with a laser medium imposes additional constraints on the way the laser can operate. It forces the laser to oscillate in modes. Modes are the various forms a wave can take within the confines of a resonator. These forms reproduce themselves in shape and phase after a round trip in the resonator. They are the particular forms that can be maintained by the resonator without excessive loss of energy. A mode is a particular discrete wavelength at which oscillation can take place. The resonator will not support oscillation at just any wavelength. The lasing medium has a gain linewidth that defines all wavelengths over which the laser can operate. But the optical resonator imposes a new constraint, permitting the laser to oscillate only at a particular set of discrete wavelengths within the gain linewidth. Each discrete wavelength or frequency of oscillation is called a mode, or mode of oscillation.

There are two types of modes, those that form along the length of the laser, called *axial* or *longitudinal modes,* and those that form in planes at right angles to the laser axis, called transverse or off-axis modes. Axial modes form because of a restriction the optical resonator places on the wavelengths that can oscillate. Only those wavelengths within the gain linewidth that are multiples of one-half the distance between the mirrors can oscillate; that is, the mirrors must be a multiple of half wavelengths apart. This condition can be written:

$$l = n \ (\lambda/2)$$

l is the spacing between the mirrors, n is a whole number, and λ is the wavelength. Since the wavelength is a very small number, n must be a large number. This condition determines the cavity resonances. With this restriction, the frequencies of operation of the laser now appear as shown in figure 5.4. This figure is a modification of figure 5.2 with the addition of the resonances. The laser will not operate with just any set of frequencies falling under the bell-shaped curve of figure 5.2; it will be restricted to the discrete set of frequencies shown in

Figure 5.4. Selection of Mode Frequencies Within the Linewidth of a NeHe Laser

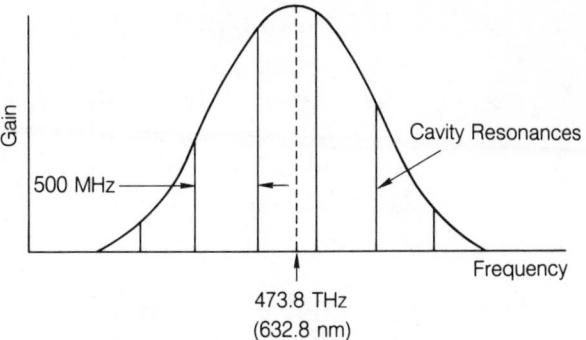

473.8 THz
(632.8 nm)

figure 5.4. Thus the modes of a laser are formed by the fact that it has gain over a certain linewidth but can oscillate only at the resonances. For the laser in the figure, these could be up to six axial modes.

The use of an optical resonator can also cause modes of oscillation to form that affect the cross-sectional intensity distribution of the laser output. These modes are called transverse or TEM modes. TEM stands for *transverse electromagnetic*. These modes are produced in as many ways as a stable optical beam can be formed by the interaction of the two mirrors. The modes represent patterns of the light beam that reproduce themselves at each mirror and therefore represent the set of possible stable operating conditions for the resonator. Some transverse laser mode patterns are shown in figure 5.5. The most common pattern is called the TEM_{oo}

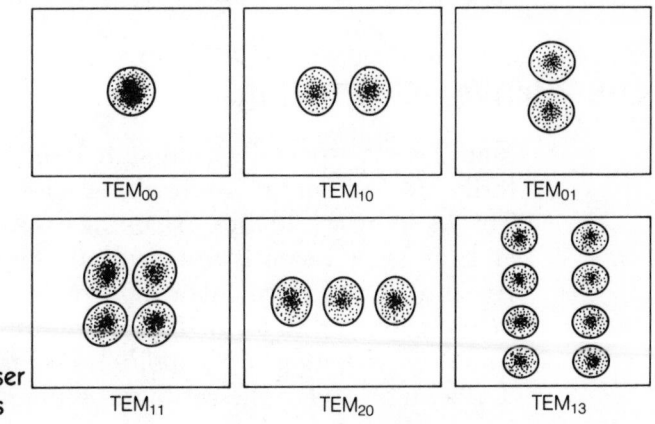

Figure 5.5. Transverse Laser Mode Patterns

mode. It produces the narrowest beam and is the way most lasers operate. It consists of a single, simple, circular spot. The other, less common modes are not as stable and are not usually used for laser operation. These "higher order" modes can be prevented from occurring by placing limiting apertures in the laser resonator.

Multimode and Single-mode Operation

With no special mode selection devices within the resonator, the laser operates by randomly selecting axial modes and jumping between them. This is referred to as *multimode* operation. Optical devices can be added to laser resonators to alter the modes of operation. One device is called an *etalon.* This is simply a flat piece of glass that is highly transmissive at the wavelength of operation and can be adjusted with respect to the laser axis. The etalon has its own set of resonant wavelengths, which are superimposed over the resonances of the laser resonator. When properly aligned, all but one axial mode will be suppressed, forcing the laser to oscillate in only one mode. This stabilizes the laser output and greatly narrows the linewidth. *Single-mode* operation is needed for a great variety of research applications that require a high degree of monochromaticity, frequency stability, or long coherence lengths. Single-mode operation can also apply to pulsed lasers but is most commonly used for continuous lasers.

Q-Switching (Q-Spoiling)

Many lasers, especially solid-state lasers, are capable of momentarily storing energy by forming an excess of excited atoms. This happens while they are being excited by the pump energy and the laser beam is not present. This extra energy will be released in the form of a high-power pulse when the laser beam is allowed to form. The operation of a laser in this way is called *Q-switching* or *Q-spoiling* and is done very simply by degrading the Q of the resonator momentarily and then restoring it. While the resonator is spoiled, lasing action ceases, but the pumping of atoms into excited states continues, forming a higher level of population inversion. The atoms stay in the excited state longer because there is no optical beam to release their energy by stimulated emission. Once the unusually high population inversion is achieved, the resonator suddenly becomes resonant (high Q), and all the energy is released as the optical beam rapidly builds up from the higher gain.

Q-switching can be achieved in many ways. It can be done simply, such as with a motor turning a circular vane with slots cut all around the edge and positioned in the resonator; the slots interrupt the beam, momentarily cutting it off and on. Modulators can also be used for Q-switching. We will discuss these later in this chapter. It is possible to Q-switch both continuous and pulsed lasers, obtaining higher power output pulses than would be obtained with normal operation. It is common to use Q-switching with ruby and neodymium solid-state lasers. We will talk about these lasers more in chapter 6.

Mode Locking (Phase Locking)

Q-switching can convert a continuously operating solid-state laser into a pulsed laser with a train of high-power pulses. Mode locking can do the same thing, but it works with more types of lasers and produces much shorter, higher-power pulses. Production of these kinds of pulses is very important for scientific research into the behavior of atoms and molecules. Mode locking does not depend on taking advantage of energy storage in the lasing medium but converts the output to a train of pulses by imposing a constant phase between all the operating axial modes. This causes them to add together, engaging in interference with each other.

A rather unexpected result occurs when the axial modes add together. Amazingly enough, the continuous beam is converted into a train of short pulses. How do the modes add up to become pulses? Look back at figure 5.4, and note that a laser can operate with any one of several axial modes whose wavelengths fall within the laser's gain curve. The wider the gain curve, the more axial modes that can be active. Without any outside influence, the laser randomly operates in multimode by jumping around and using one or more of these modes. The axial modes consist of a set of waves resonating in the cavity that are of different amplitude, each having a slightly different wavelength. As long as the waves are not in phase, that is, as long as they do not line up, adding the waves together produces nothing new or different. However, if the waves become synchronized so that their phases are fixed or locked with respect to each other, they can add together, building up by interference into a new type of wave.

The way phase-locked waves of slightly different wavelengths add up is interesting. The addition of waves is illustrated in figure 5.6. Parts a and b show two waves that are in phase but of slightly different wavelength, and part c shows the sum of the two waves. Notice that part of the time they reinforce each other and part of the time they cancel, causing the formation of a wave with lumps. When more waves with still different wavelengths are added, the result begins to look like part d. Now they cancel over a longer time and form a more concentrated lump. They begin to look like a train of pulses. Part e shows the output from a laser whose axial modes are locked in phase. The output of the laser is a train of short pulses. The more waves available to add up, the narrower the pulses. This means that when the linewidth (width of the gain curve) is wider, more axial modes are available and the pulses become narrower. The narrower the pulses, the more possible it is to study the reactions of atoms and molecules over very short periods of time.

One of the best lasers for mode locking is the dye laser, discussed near the end of chapter 6. Because of its wide linewidth, incredibly short duration pulses have been obtained. For years scientists have been working with pulses as short as a few picoseconds. Now, through the use of more refined techniques, they are working with pulses so fast that they last only a few hundred femtoseconds. Light travels only 30 micrometers (about one thousandth of an inch) in 100 femtoseconds. These are the shortest time-duration phenomena ever investigated by scientists.

Axial modes are phase locked together by placing a modulator in the resonator and switching the resonance on and off

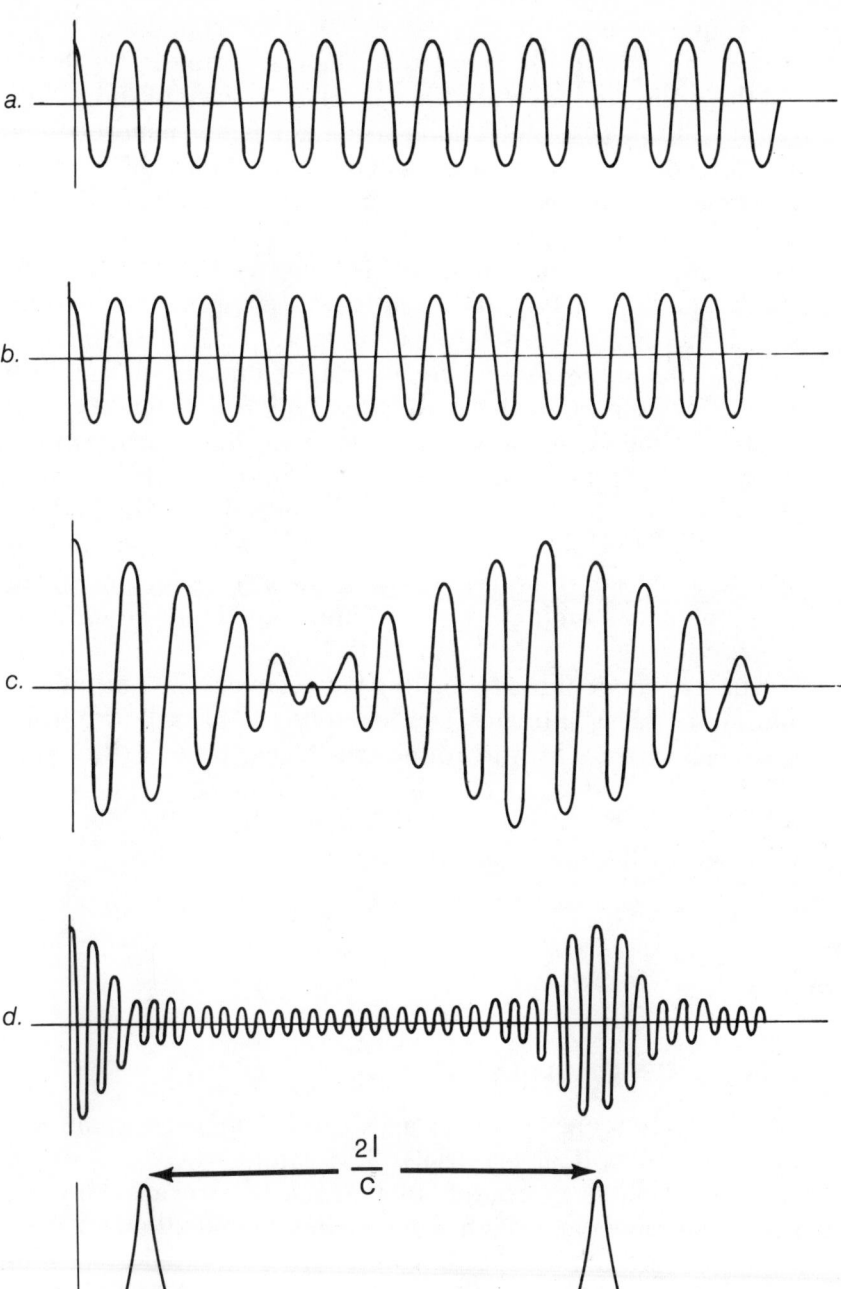

Figure 5.6. Addition of Phase-Locked Waves of Different Wavelengths

at a frequency equal to the reciprocal of the round-trip travel time of the light in the resonator. The travel time is twice the length of the resonator divided by the velocity of light. The transit time for a 1-meter-long resonator is 6.7 nanoseconds, and the modulation frequency is 150 megahertz. This method works because the axial waves add up in such a way that the spacing between pulses is equal to the resonator round-trip travel time. Thus, the modulator is being switched on and off at a frequency that tends to support the production of the pulses, thereby locking the axial modes in phase. Pulsed lasers can also be mode locked by placing a liquid cell called a *saturable absorber* in the resonator. This is a special type of substance that is opaque to low-intensity light but becomes transparent for high-intensity light. The saturable absorber favors the production of the pulses by permitting the higher-level energy in the optical beam to oscillate and stopping the lower-energy continuous output. It does in effect create a resonator with a high Q for pulses and a low Q for continuous operation.

The net effect of mode locking is to convert the output of a continuous laser into a series or train of ultrashort pulses separated in time by the round-trip travel time of the laser resonator. Pulsed lasers can also be mode locked. This results in pulses of shorter time duration and higher power than would normally be produced.

Modulation

ACOUSTO-OPTIC MODULATION

Very-high-frequency sound waves can be imposed upon certain types of optical materials. The sound waves are much higher in frequency (around 500 megahertz) than anything the human ear can hear, which is approximately 20 kilohertz. The sound waves, called *acoustic waves*, are generated electronically and applied to a thin layer of material called a *transducer*, which converts the electrical signal into a vibration. The transducer is a piezoelectric crystal that converts electrical signals into mechanical movement. When the acoustic waves travel in the lattice of certain crystals, they change the index of refraction. This makes the crystal appear to the light wave like a diffraction grating or prism, which will deflect the light beam, changing its direction. Crystals that can properly carry the acoustic waves and transmit the laser beams are called *acousto-optic* crystals.

This ability to electronically deflect the light beam can be put to many very useful purposes. The very-high-frequency signal used to modulate the index of refraction can itself be modulated with a lower-frequency signal. The lower-frequency signal is used to rapidly turn the acoustic action on and off, to change the deflection of the beam, and thus to modulate the laser output. Acousto-optic devices can be used to modulate lasers with signals at frequencies from a few hertz up to 100 megahertz.

Acousto-optic modulators are used for Q-switching, mode locking, beam deflection, and beam modulation. For Q-switching, the crystal is modulated at the frequency that will produce the highest-power output pulses. Beam deflection and beam scanning are used for laser-controlled printers for computers, for audio and video disk playback systems, and for optical information disks for computer storage. Beam modulation can be used for data links and communications, while mode locking is commonly used in research projects. Some of these applications will be discussed in chapter 7.

ELECTRO-OPTIC MODULATION

Information can also be encoded in laser beams by *electro-optic* modulation. For some optical crystals, a change in index of refraction can be directly induced by applying an electric field. By modulating the application of the electric field with a signal, the information in the signal can be encoded in the laser beam. Depending on the design of the system for modulating the beam, the signal can be encoded by changing the phase, the amplitude, or the polarization of the laser beam. The laser beam travels down one axis of the crystal, and the electric field is applied perpendicular to the direction of travel of the laser beam. Electro-optic modulation can be applied either to laser beams traveling through the atmosphere or to those traveling in fiber-optic cables.

The faster the electric field can be changed, the more information can be encoded on the laser beam. The rate at which the electric field can be changed is called the *modulation frequency*. There are many factors in the properties of electro-optic crystals and in the way the electric field is applied to the crystal that limit the modulation frequency that can be achieved. Scientists have been working for many years to improve on materials and techniques for design of electro-optic systems. Their goal is to increase modulation frequencies so that laser beams can carry more information. Commercial systems operate well into the 100 megahertz range, and laboratory systems are being developed that make it possible to move into the gigahertz range.

Electro-optic modulators are used in laser communications systems for fast, practical, high-capacity information transmission. The military likes optical communication because light beams travel directly from one location to another, making it difficult for the enemy to listen in. Electro-optical modulators are used in systems for high-speed recording of images or data on photographic film. They are also used in audio and video disk recorders for high fidelity and television playback. Some very-high-speed, high-density optical storage systems for computers are based on electro-optic modulation. Optical disks can contain much more information and therefore are more practical than the magnetic disks of the same size that are now in common use.

Summary of Laser Operation

In summary, laser operation is made possible because a broadband, nonmonochromatic source of energy can be used to pump the electrons of atoms into higher energy levels. These decay into a quasi-stable, long-lifetime level where passing photons of light stimulate them to emit more light energy in unison, increasing the intensity of the optical field. Reflecting mirrors at each end of the lasing material feed the optical field back on itself, until the field is so intense that it extracts most of the energy available from the excited electrons. Before the laser was developed, all light was emitted by electrons that gained their energy as a result of the temperature of the material, such as the hot tungsten filament in a light bulb. For a filament to produce optical fields at intensities comparable to those from lasers, temperatures of millions of degrees would have been required. Even if that had been possible, the light would not have been coherent or monochromatic.

Lasers that Don't Work the Way We Just Said They Do

There are several types of lasers that achieve population inversion through methods different from those just described. These are semiconductor laser diodes, chemical lasers, gas-dynamic lasers, and free-electron lasers. The last three types of lasers are complex, usually quite large, and difficult to build. They are used for special purposes, mostly for research and for the military. Gas-dynamic lasers are not available commercially, but a few chemical lasers are available. The semiconductor laser diode is the most common type of laser in use. It is very simple and can be very inexpensive. It is, therefore, a very important type of laser.

SEMICONDUCTOR DIODE LASERS

Semiconductors are the substances that are used to form diodes, transistors, and integrated circuits used in all modern electronic equipment. Semiconductors are combinations of chemical elements that form a material that falls somewhere between conductors that freely permit the flow of electrons (like metals) and nonconductors that allow no electron flow (like plastic or wood). The flow of current in a semiconductor depends on impurities that are purposely placed in the semiconductor and on the combination of semiconductors with different impurities placed together to form semiconductor diodes. In this section we will talk about how semiconductors can be used to make lasers and how these lasers work. Chapter 6 will present more details on the advantages, disadvantages, and uses of *semiconductor diode lasers*. From now on, we will shorten the name to *diode lasers*.

Diode lasers are similar to other lasers in that they cause electrons to emit light by transitioning to and from different energy levels. However, in the semiconductor these energy levels are really energy bands associated with the cumulative effects of the outer electrons of the atoms that make up the semiconductor material, as shown in figure 5.7. There are two bands and a gap between them. The lower band is called the *valence band*. When an electron is in the valence band, it cannot move about and contribute to the conductivity of the semiconductor. It is attached to an atom or group of atoms. The upper band is known as the *conduction band*. When an electron has enough energy, it can jump the band gap and

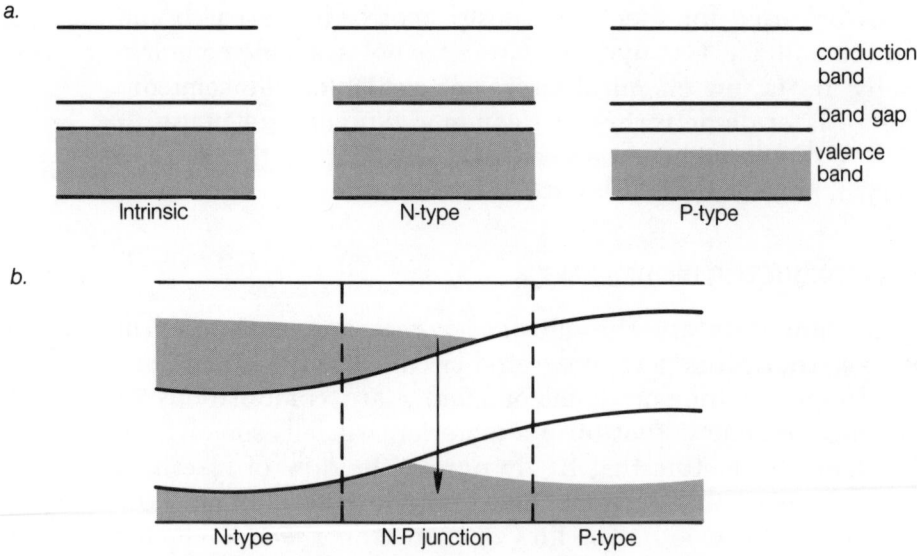

Figure 5.7. Energy Bands in Semiconductors

move into the conduction band, where it can move around and contribute to the conductivity of the material. In an *intrinsic semiconductor*—that is, one that has no impurities—the valence band is almost full and there are very few electrons in the conduction band. Certain types of impurities can be added to semiconductors creating an excess of electrons in the valence band, thus forcing more electrons into the conduction band. Other types of impurities force openings in the valence bands, resulting in an absence of electrons that would normally be there. These absences are called *holes*. When impurities are added, the semiconductor is called *doped*. When an impurity is added to create excess electrons, the semiconductor is called *N-type*, and when an impurity is added to create a lack of electrons it is called *P-type*.

A most important thing occurs when N-type and P-type semiconductors are brought together. They form what is called a *P-N junction*. The P-N junction is the basis for all modern electronics. It is used to form diodes, transistors, integrated circuits, diode lasers, and light detectors. The formation of a junction is shown in figure 5.7–*b*. The region of the junction is very narrow, and the excess electrons on the N side tend to migrate across the junction toward the P side. At the same time, there is a tendency for the holes to migrate across the junction in the other direction. Once there is an energy gap between the electrons and holes, the electrons want to move into the holes but must transition the energy gap to do so. The energy gap in semiconductors is usually called the *bandgap*. When the electron transitions, it emits light in the same way as when transitioning energy levels in an atom. The population inversion takes place between the electrons in the conduction band of the N side and the holes in the valence band on the P side. Both spontaneous and stimulated emission can be the result. The frequency of the emitted light is proportional to the bandgap, just as the frequency of emitted light from an atom is proportional to the change in energy level. The output wavelength of diode lasers can be controlled by using semiconductors with different bandgaps. When a power supply is added, it shifts the energy-band levels with respect to each other, as shown in figure 5.7–*b*, and forces more electrons to move from the N side to the P side and more holes to move from the P side to the N side. Referred to as *carrier injection,* this process increases the population inversion, making more electrons available for transitions. The transitions result in recombination of the carriers, which occurs in the narrow region between the N and P materials called the *recombination region.*

Like other lasers, the diode laser must have a resonator in order to oscillate and develop laser-type output. Diodes emit light spontaneously without a resonator. In this case the light is incoherent and uncollimated. If there is no resonator, the diode is called a *light-emitting diode,* or LED. Many digital watches and pocket calculators use LED displays. The resonator for a diode laser is formed by the material that surrounds the P-N junction and by the end planes of the material perpendicular to the junction, which are cut, polished, and sometimes optically coated. Once the resonator is available, stimulated emission builds up, and the diode output becomes more coherent, more collimated, and more powerful. The diode lases.

The P-N junction is only one type, the simplest type of junction for diode lasers. It is referred to as a *homojunction* since the semiconductor on both sides is the same. Scientists have developed other types of junctions to create diode lasers for special purposes and to improve their output power and efficiency. Three types of junctions are shown in figure 5.8. Although there are other types of junctions and different structures for the semiconductor materials to form diode lasers with special capabilities, they are too complex to go into here.

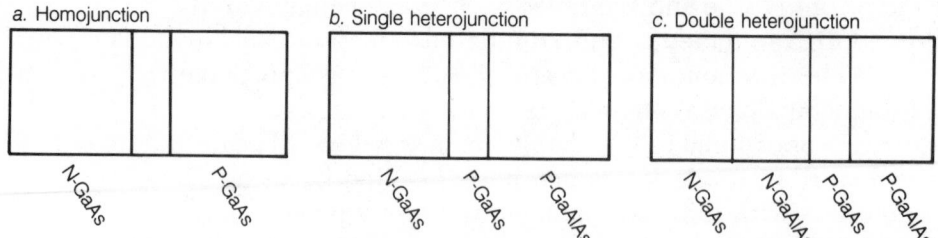

Figure 5.8. Different Types of Junctions in Diode Lasers

A common type of material for diode lasers is gallium arsenide (GaAs). Another type of material similar to GaAs but with a slightly different band gap is gallium aluminum arsenide (GaAlAs). Placing P-type GaAs between P-type GaAlAs and N-type GaAs forms what is called a *single heterojunction*, shown in figure 5.8–*b*. For rather complex reasons, the single-heterojunction lasers, by bringing together substances with different bandgaps, have better operating properties than homojunctions. Lasers with single heterojunctions are much better for pulsed operation. They emit the highest-peak power when operated with pulsed electronic power supplies.

The *double-heterojunction* laser is formed as shown in figure 5.8–*c*. This junction sandwiches a P-GaAs–to–N-GaAlAs junction between P-GaAlAs and N-GaAs. This type of diode laser works much better for continuous-output operation and is most commonly used in fiber-optics communications. The double-heterojunction laser requires less power and can be modulated at very high frequencies. The capabilities and applications of diode lasers are further discussed in the next chapter.

CHEMICAL LASERS

Chemical lasers are a class of lasers that attain population inversion through the chemical reaction of two atomic species to form vibrationally excited molecules. The term *chemical laser* refers to a set of different lasers. Among these are:

- Iodine (I)
- Hydrogen fluoride (HF)
- Deuterium fluoride (DF)
- Hydrogen chloride (HCl)
- Hydrogen bromide (HBr)
- Carbon monoxide (CO)

The most common chemical lasers are the first three. All chemical lasers emit in the infrared region of the spectrum. They can be operated both pulsed and continuously, and their output can be very powerful. HF lasers emit over a range of wavelengths from 2.6 to 3.3 micrometers, and DF lasers emit over a range from 3.5 to 4.2 micrometers.

In an HF chemical laser, molecular hydrogen, H_2, or molecules containing a great deal of hydrogen (such as various hydrocarbons), are combined with molecular fluorine, F_2, or a molecule containing fluorine (such as sulphur hexafluoride, SF_6) by rapidly flowing them into a reaction chamber where they form HF molecules. If SF_6 is used, fluorine atoms must be broken off by an electric discharge. The important fact about these types of reactions is that the newly formed molecules, in this case HF, are vibrationally excited. If a resonator is provided, they emit infrared light by spontaneous and stimulated emission. Also important is the fact that nearly all the molecules formed are vibrationally excited. This results in a very high population inversion capable of producing high gain and opens up the possibility of producing very high output power. Special research chemical lasers have produced two megawatts continuously and are being designed to produce five megawatts. Because these power levels are high enough to damage objects over considerable distances, they are of interest to the military. Chemical lasers are unique in that their power comes from the chemical reactions, and electronic discharges are not required. Chemical lasers, as well as gas-dynamic lasers, are good candidates for a space-based defense system. The lasers we have today are far from adequate for that use. Considerable research and development will be required not only on the laser devices themselves but on identification and pointing systems used to pick out and identify targets and to properly direct the laser energy. All these complex systems must be perfected before space-based laser systems can be built.

GAS-DYNAMIC LASERS

Certain types of molecules, most commonly carbon monoxide or carbon dioxide, can be compressed to very high pressure and allowed to flow through a very narrow nozzle, much like a miniature jet engine. When the gas rapidly cools and expands after passing through the nozzle, the molecules are left vibrationally excited. As in chemical lasers, a very high population inversion can be formed, resulting in high gain and very high output powers. A gas-dynamic carbon dioxide laser is the most powerful type of laser with continuous output ever built. It is discussed in the next chapter.

FREE-ELECTRON LASERS

The free-electron laser (FEL) produces optical radiation by decelerating high-speed electrons from a particle accelerator through static magnetic fields. These magnetic fields alternate in polarity down the laser's longitudinal axis. The principle of the FEL is based on the fact that electrons do not have to be associated with atoms in order to emit light. Rapid deceleration of electrons in electric or magnetic fields causes them to change kinetic energy, thereby radiating optical energy. The frequency of the photons emitted is proportional to the change in energy of the electrons.

The advantages of such a laser are many. The output wavelength is controllable; in theory, it is tunable over the entire electromagnetic spectrum. The tunability arises from the fact that the output wavelength of an FEL depends on its design and operating parameters and not, as with other lasers, on the naturally occurring electron energy levels in atoms. The FEL is a potential source of tunable wavelength radiation at wavelengths where no sources currently exist. The regions of the electromagnetic spectrum where optical sources are needed are wavelengths shorter than 100 nanometers and wavelengths in the range from 25 to 1,000 micrometers. The FEL is best suited to produce long wavelengths. Shorter wavelength production requires much higher initial electron energy from the accelerator and stronger static magnetic fields that alternate over shorter distances. Typical output wavelengths of FELs are longer than 3 micrometers, in the infrared region of the spectrum. Output has been obtained at 1.6 micrometers and 630 nanometers, but the output powers are very low, less than 100 microwatts in the visible range. Since they require high-energy electron accelerators and large magnets, present-day FELs are large, heavy, and very inefficient. However, in the long run, the free-electron laser is potentially a very important source of laser output, in a very broad range from the ultraviolet to the far infrared regions of the spectrum. This laser is especially suitable for research.

This chapter has presented fundamental concepts that are necessary to understand how lasers work, why there are so many different kinds of lasers, and how various optical devices are employed with lasers to expand their usefulness.

The operation of most lasers is based on transitions of electrons between electron energy levels in atoms. Most of these lasers operate at a discrete wavelength or a discrete set of wavelengths. Many lasers operate from vibrational or rotational energy transitions in molecules. Some of these lasers have selectable output wavelengths. Several types of lasers have unique ways of achieving population inversion. These include carrier injection in semiconductors, chemical reactions, expansion of gasses in nozzles, and decelerating electrons in magnetic fields.

In the next chapter, we will look in more detail at many of the most popular types of lasers, how they are built, the characteristics of their output, and how they are used.

6

Common Types of Lasers and Their Characteristics

Now that we have looked at the fundamental concepts behind laser operation, let us look at some of the most common lasers. This chapter presents specific information on how each type of laser operates, and on the types of output, the power levels, linewidths, reliability, and costs of the lasers. A familiarity with the capabilities and limitations of most types of lasers allows an understanding of how each type of laser can and cannot be used within the context of the many fascinating and valuable ways that lasers are being applied to technology.

Helium-Neon (HeNe) Lasers

One of the most common lasers is the helium-neon laser. It is a reliable, easy-to-build, low-cost source of continuous light in either the visible red or the infrared regions of the spectrum. The lasing medium is a mixture of helium and neon gas sealed in a glass tube at low pressure. The gas is excited by a few milliamperes of direct current applied at several kilovolts. The electrons passing through the gas collide with both the helium and neon atoms, exciting their electrons. Most of the excitation energy is collected by the helium atoms and transferred by collisions to the neon atoms. This occurs because helium has an excited state that is very close to the level of the excited

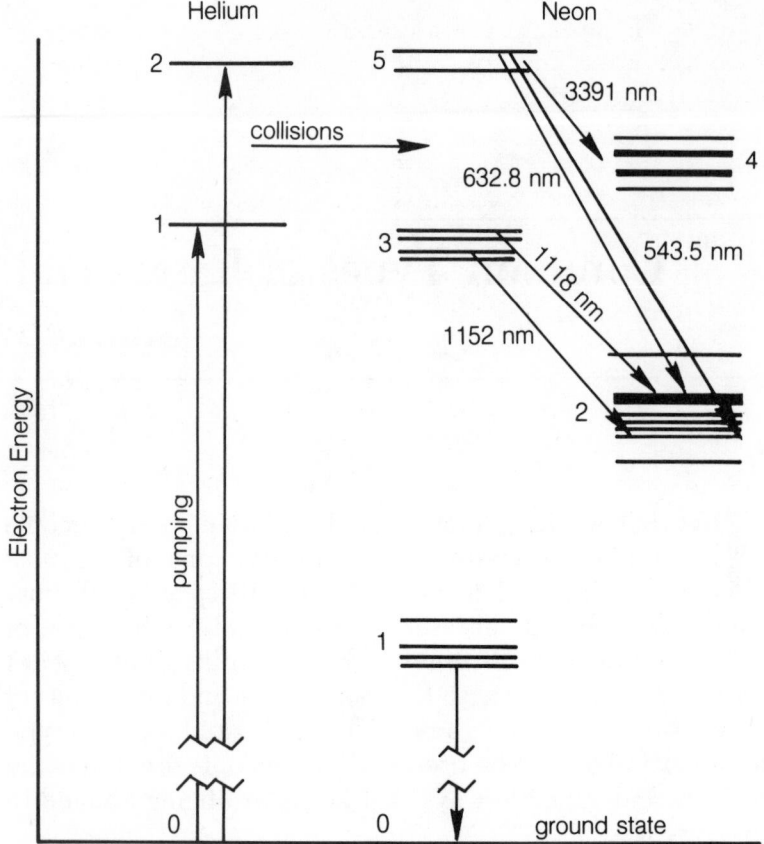

Figure 6.1. Energy Levels in Helium and Neon Atoms

state that leads to laser action in neon. The energy levels for helium and neon are shown in figure 6.1. As shown in the figure, neon can operate with four different energy-level transitions. These produce output at 543.5-, 632.8-, 1,118-, 1,152-, and 3,391-nanometer wavelengths. The most common use is at 632.8 nanometers, which is red visible light. Most HeNe lasers are built to produce the red output, although many are being used at 1,152 nanometers. The first gas laser built by Ali Javan and his associates was a continuous-wave HeNe laser. It operated at 1,152 nanometers. The other wavelengths are seldom used. This laser is seldom used as a pulsed laser but can be mode locked to produce a pulse train output.

The HeNe laser is in common use because it is an inexpensive source of continuous, highly collimated, very monochromatic red light. The output is stable and of a very narrow linewidth. These lasers are reliable, reasonably portable, and cost as little as a few hundred dollars. They can last for up to 20,000 hours. The power output is from fractions of a milliwatt to 50 milliwatts. It is not practical to build high-power HeNe lasers, since they are able to convert at best 0.1 percent of the input excitation energy to output light energy; in other words, they are rather inefficient. A large percentage of their energy is given up in the form of fluorescent light and heat. Because of their stability, narrow linewidth, and continuous collimated output, they have many practical applications.

The power output can be held constant to within about 1 percent. This is called the *amplitude stability.* The frequency of the output of a HeNe laser operating single mode and held at constant temperature will vary as little as 2 parts in a billion while operating over periods of several minutes—a rather impressive *frequency stability.* The width of the gain curve is 1.4 gigahertz, which corresponds to approximately 0.002 nanometer (2 picometers) in wavelength while emitting red light. Several axial modes fall under the gain curve (similar to that shown in Figure 5.4). The laser can operate multimode or single mode. In single mode it is highly monochromatic.

The most common use of HeNe lasers is for alignment and positioning. Almost every optical and laser research laboratory in the world has a HeNe laser. They can be used to prealign optics for almost any kind of optical experiment. They are especially useful for aligning other kinds of lasers where the output is not visible or where it is pulsed and therefore not possible to use for alignment; most lasers will not operate until their optics are closely aligned. HeNe lasers are also commonly used in the construction industry, where their beams may be scanned back and forth to accurately locate planes or lines. They are used to locate paths for machine tools and to guide grading by earth-moving equipment. The laser assures that the tools or equipment move in straight or level lines and into correct locations.

HeNe lasers are also used for reading and scanning. The beams can be scanned through the use of rotating mirrors,

(Photograph courtesy of Melles Griot Laser Products, San Marcos, California)

These helium-neon laser tubes are simple and compact. The resonator mirrors are built into the gas tube, eliminating the need for Brewster windows. The metal cylinder at one end of the tube forms the cathode, called a hollow cathode, while the anode is at the other end. Also shown is a power supply.

acousto-optics, or other optical means. The most popular application is in automatic checkout stands in supermarkets. The laser sends out its beam, which is rapidly scanned in many directions to illuminate the coded bar chart on purchased goods. Detectors inside the device pick up the light reflected from the bar chart and determine from the way the bars reflect the light which product it is. A computer provides the rest of the needed information such as type of goods and price.

HeNe lasers are used in very-high-speed computer printers. Since they have good coherence and are low cost, they are commonly used in holography. We will talk in more detail about both of these applications in chapter 7. They can be used in interferometers to measure the surface contours of objects and are used to measure the dimensions of objects while the objects are being manufactured. Another fascinating application—the ring laser gyroscope that can be used to measure the location, speed, and acceleration of aircraft and spacecraft—will also be discussed in chapter 7.

Carbon Dioxide (CO_2) Lasers

The carbon dioxide (CO_2) laser is also a gas laser but very different from the HeNe laser. The emission from the CO_2 laser is based on changes in vibrational energy levels in the carbon dioxide molecule. This very common molecule, abundant in the atmosphere, is used to make one of the most versatile, useful, and efficient lasers known. It emits in the far-infrared region of the spectrum at a number of wavelengths ranging from 9 to 11 micrometers (9,000 to 11,000 nanometers). The most powerful and most commonly used emission is at 10.6 micrometers. It can be operated continuously, in single pulses or in pulse trains. The pulsed modes of operation are desirable when higher powers are needed. CO_2 lasers are highly efficient. They can convert from 5 percent to 20 percent of the input excitation energy into output infrared light. The efficiency depends on the type of laser and the mode of operation. There are six distinct types of CO_2 lasers, which can range in output power from a few watts continuously for simple laboratory lasers to a few megawatts continuously for some military lasers. CO_2 lasers can be made to be powerful enough to destroy missiles or to weld metals. The most powerful continuous-output laser built is a CO_2 laser. Continuous CO_2 lasers do not last as long as HeNe lasers: they remain operative only up to a few thousand hours. They also cost more, from about $2,500 for the simplest ones up to $1 million for the powerful ones used for metal welding, cutting, and similar applications.

A vibrational energy level diagram for carbon dioxide is shown in figure 6.2. This diagram is similar to those for the electronic energy levels of atoms except that the energy is present in the molecule in a different form, that of molecular vibration. Electrons surrounding atoms gain and lose energy

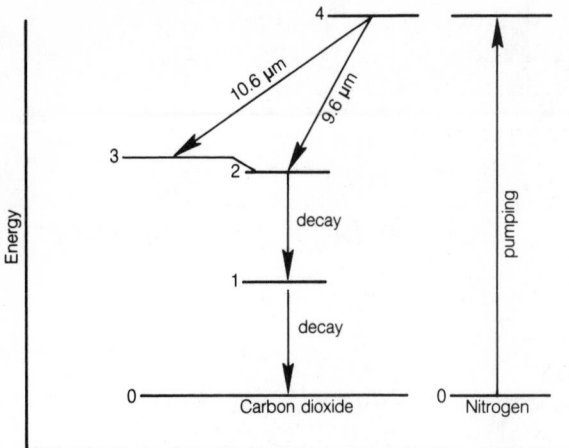

Figure 6.2. Energy Levels for Carbon Dioxide Molecules

by changing electron energy levels that naturally occur in atoms. Molecules, combinations of atoms bound together by bonding forces of their outer electrons, can gain and lose other forms of energy such as vibrational or rotational energy. The carbon and oxygen atoms vibrate against one another like small objects held together with springs, giving rise to vibrational energy levels. The CO_2 laser must also contain nitrogen and helium in order to work. The nitrogen molecule (two nitrogen atoms bound together) absorbs energy from the excitation electrical discharge. It can transfer this energy to the carbon dioxide molecule by collisions because it has a vibrational energy level very near to that of carbon dioxide. The CO_2 molecule is excited to level 4 and can transition by stimulated emission to either level 2 or level 3. Level 3 works best and puts out the 10.6-micrometer light. Level 2 puts out 9.6-micrometer light. The level used depends on the resonance wavelength of the optical resonator. Level 3 will decay in a nonradiative manner to level 2. To complete the process, level 2 must decay to level 0 or the population inversion will not be maintained. If the molecules stay in level 2, eventually not enough excitation will be provided to keep level 4 more populated than levels 2 and 3. For this reason, helium is mixed with the carbon dioxide and nitrogen. Collisions with the helium atoms help the carbon dioxide molecules to drop from level 2 to level 1 and to level 0. In the figure, levels 1 and 2, 3 and 4 are shown in different columns because they represent different types of vibration.

CO_2 lasers can be built in six different ways, depending on what type and how much power is desired. Generally, the more power desired, the more complex and expensive the laser. We will discuss each type of laser, how it is built, and how it can be used.

SEALED-TUBE CO_2 LASERS

This is the simplest type of CO_2 laser, but because of certain drawbacks, it is not commonly used. It consists of a sealed tube of CO_2, N_2, and He that is excited by an electrical discharge applied along the length of the tube, just as is done with HeNe lasers. This is sometimes referred to as an *axial discharge*. A simplified drawing of a sealed-tube CO_2 laser is shown in figure 6.3–*a*. The discharge is struck between the electrodes at each end of the laser. Some CO_2 lasers are excited by *transverse discharges*, in which the electrical excitation is applied perpendicularly to the length of the laser. When the gases are excited by the electrical discharge, the CO_2 molecules tend to disassociate into carbon monoxide (CO) molecules; this destroys the laser's ability to perform. In the sealed-tube laser, substances are provided to help the CO_2 molecules reform. In other types of lasers, the gases are replaced or restored and recirculated. Sealed-tube CO_2 lasers are limited to maximum output powers of approximately 50 watts continuously; otherwise, they have to be built too long to be practical.

AXIAL-FLOWING GAS CO_2 LASERS

In axial-flow CO_2 lasers, the problem of replenishing the gases is solved by flowing fresh gas down the length of the laser as illustrated in figure 6.3–*b*. Since the gases are at low pressure, the consumption of gas is not excessive. These lasers can produce more power than sealed-tube lasers. A 1-meter-long axial flow CO_2 laser can produce from 40 to 80 watts continuously. Longer lasers and especially those that fold the tube

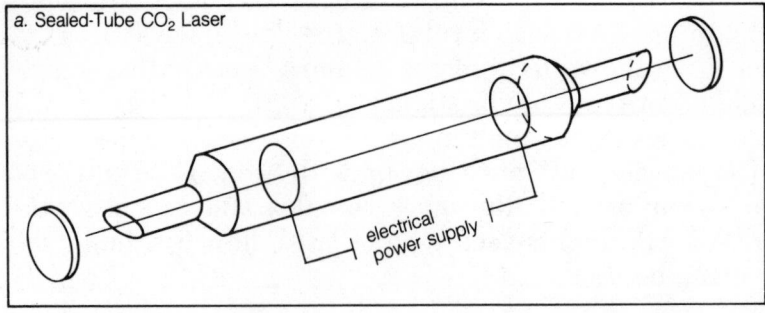

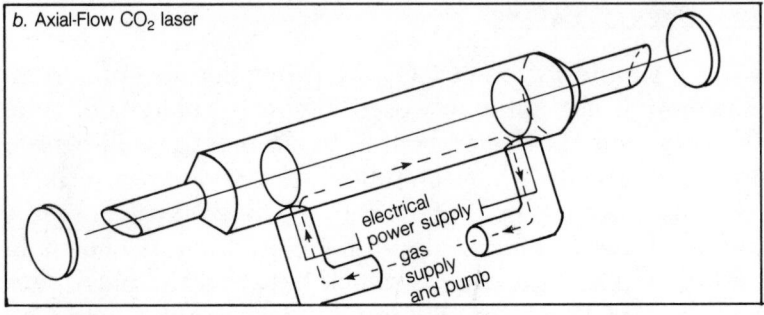

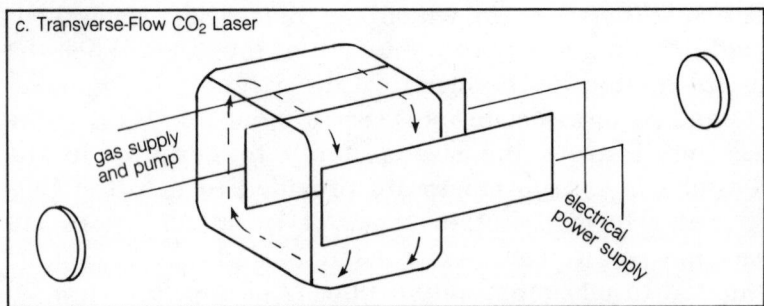

Figure 6.3. Schematic Diagrams of Various Types of Carbon Dioxide Lasers

segments so that the laser does not have to be so long can provide up to two kilowatts of continuous output. Higher powers than this are impractical with this type of laser.

TRANSVERSE-FLOW CO₂ LASERS

Much higher power can be obtained if the gases are flowed transversely (perpendicularly) to the laser length or axis. In

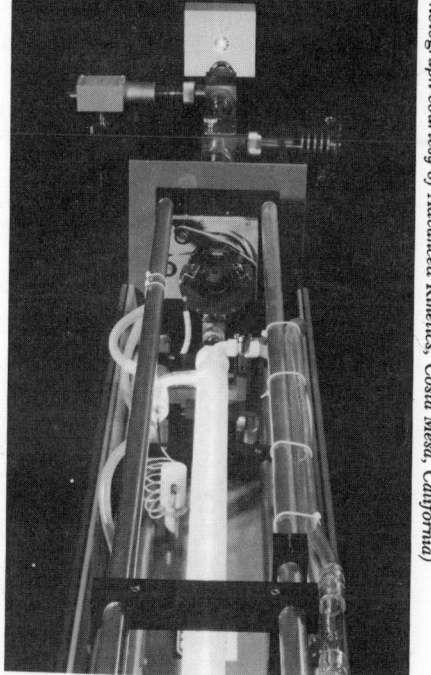

(Photograph courtesy of Advanced Kinetics, Costa Mesa, California)

An axial-flow CO_2 laser in operation. The bright area in *center left* is the discharge tube. *Above* and *below* are low-thermal-expansion-coefficient rods. Tubes adjacent to these rods bring cooling water to flow around the discharge tube. *Left of center* are devices for aligning the resonator optics. The bright spot at *left center* is a plate being heated by the laser's output.

addition, the electrical discharge can provide much more effective excitation of the gases when it is applied transversely to the axis and transversely to the flow. Since both of these methods are much more difficult to design, build, and manufacture, these lasers are much more complex and much larger. However, the output power they provide makes them useful for a whole new class of applications, such as metal, plastic, cloth, and rubber processing. They can cut, drill, shape, bend, and heat-treat better and faster than any device previously invented. A simplified diagram of a transverse-flow laser is given in figure 6.3–c.

The transverse-flow laser is more powerful than the sealed-tube and axial-flow lasers because the gas flow quickly removes excess heat from the active area and replenishes the gases. Because of the transverse discharge, the gases can be used at higher pressures than the gases in the laser types previously discussed. Since the gas is at higher pressures, the lasers do not have to be as long to achieve higher power. This type of laser is the highest-power laser commercially available at the present time. These lasers can produce up to 25 kilowatts continuously and therefore can be used to weld, cut, drill, and shape metals. The military and various research laboratories have built even more powerful versions of this laser.

GAS-DYNAMIC CO_2 LASERS

This laser is similar to the transverse-flow laser in that the gases flow in a direction perpendicular to the laser axis. But they are very different in the way the gases are excited into population inversion. In this laser there is no electrical excitation. The gas is heated and compressed to a high pressure. Then it is allowed to flow through a narrow nozzle that extends along the length and adjacent to the laser axis. The gases pass through the nozzle at supersonic speeds, rapidly expanding and cooling, causing the CO_2 molecules to be excited into the vibrational energy levels necessary for laser action.

The gas-dynamic laser operates best with an unstable optical resonator, which does a good job of extracting energy from the excited gases. This type of laser can generate the most power continuously available from any kind of laser. Power levels of 100 kilowatts or more have been reached and we can only suppose, in view of all the interest in laser ballistic missile defense, that even more powerful models may be built. Because of its complexity and size, this type of laser is not available commercially, nor does it appear to be practical on the battle-field, although it definitely has enough power to destroy objects at considerable distances.

WAVEGUIDE CO_2 LASERS

The waveguide laser is a special design suitable for reliable low-power, continuous output. It is smaller and less expensive than most of the other types of lasers. Instead of an optical resonator, the optical field is confined by a very narrow dielectric waveguide similar to that used to contain microwaves: The dielectric is highly reflective to the infrared light generated by the CO_2 molecules. The waveguide extends along the length of the laser axis containing the light beam; it extends along the sides instead of at the ends, as an optical resonator would. The small diameter of the waveguide makes it possible to get more output power from the laser with less input power; that is, the efficiency is increased. This type of laser works well for applications requiring from fractions of a watt to 50 watts of output power at or near 10.6 micrometers. It is commonly available commercially and can be as inexpensive as a few thousand dollars. Because of the high efficiency of CO_2 laser action, these lasers produce relatively high powers from smaller power supplies. A power supply is the necessary electrical device to convert 110-, 220-, or 440-volt AC line current into the current and voltages required to excite the atoms or molecules in laser gases.

TRANSVERSELY EXCITED ATMOSPHERIC (TEA) CO_2 LASERS

For continuously operating lasers, it is not possible to operate the laser unless the gases are maintained at a partial vacuum; that is, at a lower pressure than that of the atmosphere. When the pressure is increased, it is no longer possible to keep up the electrical excitation with any degree of stability. It is desirable, however, to operate at higher pressures because there are more atoms or molecules in the gas and higher output powers can be obtained. It is possible to electrically excite higher-pressure gases that are near, at, or above atmospheric pressure by using sudden powerful pulses of electrical energy. These pulses are usually created by charging large capacitors with DC power supplies and discharging the capacitors across the laser gas chamber. In this case, the electrical discharge is applied in a direction that is perpendicular to the laser axis in the same manner as with the transverse-flow lasers.

The TEA laser is desirable because intense pulses of light energy can be obtained from relatively small lasers. The gas is also at higher pressure, causing the linewidth to broaden due to collisions between the molecules. This linewidth broadening permits the laser's axial modes to be mode locked, producing trains of very intense pulses whose duration may be as short as a nanosecond. TEA lasers are commercially available.

For many of the CO_2 laser designs that have been described, gases other than CO_2 can be provided to obtain laser output at different wavelengths, although the optics have to be changed to keep the resonators at a low-loss level. A few examples are: carbon monoxide, which emits at 5.0 micrometers; hydrogen fluoride, which emits at many wavelengths between 2.6 and 3.3 micrometers; deuterium fluoride, which emits at wavelengths between 3.5 and 4.2 micrometers; and excimers (to be discussed later in this chapter), whose output is in the ultraviolet region of the spectrum.

As will be seen in chapter 7, the CO_2 laser has a very large number of important and very useful applications. The high-power lasers are used for welding, cutting, drilling, heat-treating, and shaping materials in industry, and as potential weapons for the military. They can also be used in medical applications, such as surgery and treatment of tumors. A focused CO_2 light beam can cleanly cut living tissue and result in less bleeding than a scalpel. Lower-power lasers are used in spectroscopy research to probe the properties of molecules, by looking for how they absorb, reflect, or reemit the 10.6-micrometer light, and in photochemistry research to see how chemical reactions are affected by the light. The military uses low-power CO_2 lasers to locate and measure the distance to objects on the battlefield. This is called *range finding*. The output from CO_2 lasers can also be used, much like radar, to locate and identify objects. They are used in remote sensing to pick up reflections from the atmosphere and to analyze them to determine the presence of various types of chemicals or pollutants. Some of these applications will be discussed in more detail in chapter 7.

Neodymium (Nd) Lasers

The neodymium (Nd) laser is a solid-state laser, which means that the lasing medium is a solid, usually consisting of an optical crystal with atoms of the lasing material embedded in

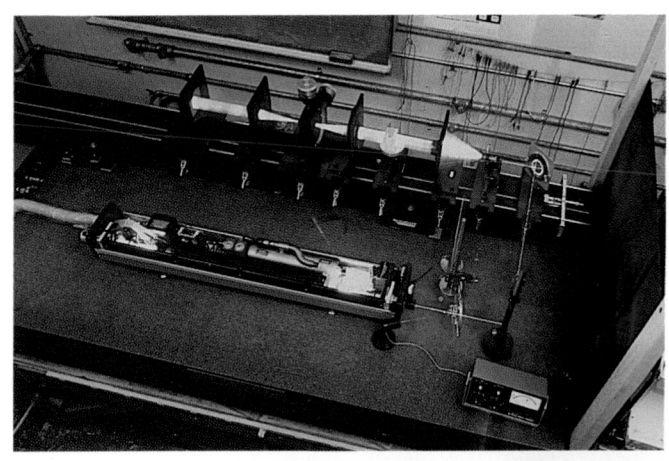

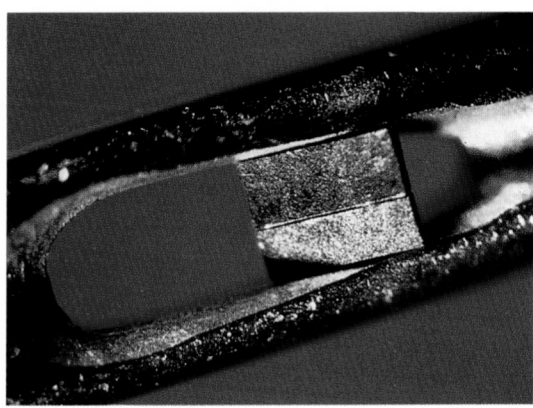

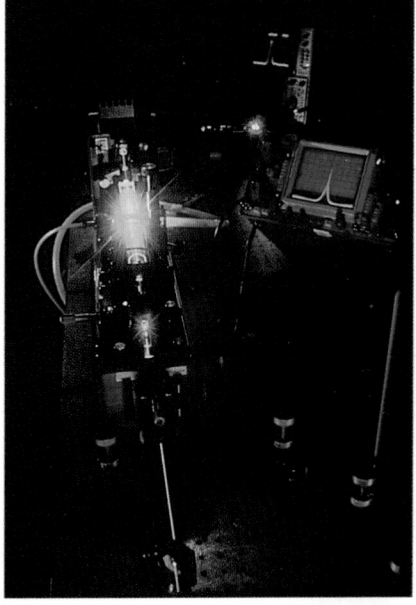

(*Top left*) An axial-flow CO_2 laser in operation. The bright area in *center left* is the discharge tube. *Above* and *below* are low-thermal-expansion-coefficient rods. Tubes adjacent to these rods bring cooling water to flow around the discharge tube. *Left of center* are devices for aligning the resonator optics. The bright spot at *left center* is a plate being heated by the laser's output. (*Top right*) An argon-ion laser in operation. The argon laser is to the *lower left of center* and the discharge can been seen at each end of the long rectangular enclosure. (*Bottom left*) This semiconductor diode laser is small enough to fit within the eye of an ordinary needle. (*Bottom right*) A scientist is tuning an experimental dye laser which produces visible light in the yellow region of the spectrum. The pumping energy from another laser is entering from the *lower left*. The laser beam originates at the bright spot at the *lower left of center* and is reflected by mirrors.

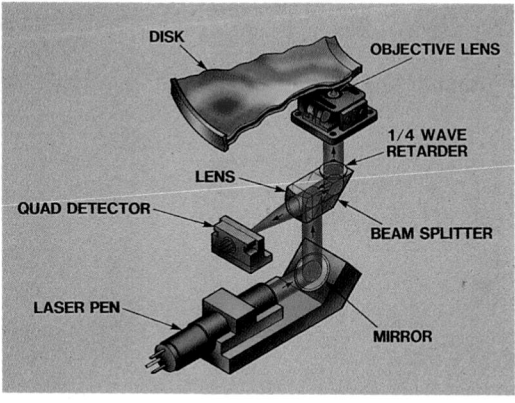

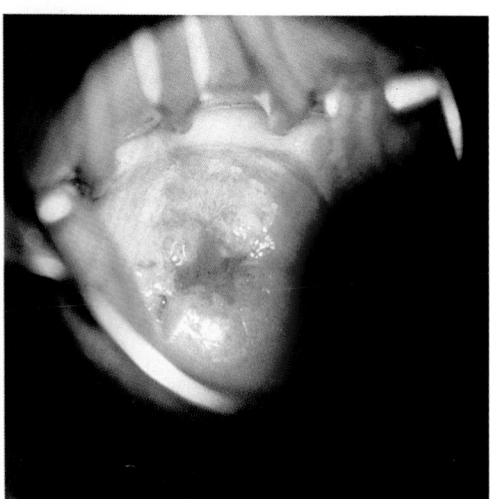

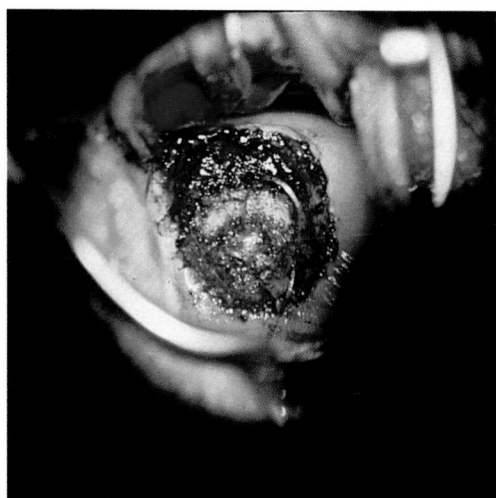

(*Top left*) A ring-laser gyroscope contains an optical system consisting of six mirrors forming three square resonators with their planes perpendicular to each other. (*Top right*) The heart of the digital optical disk drive is the carriage light assembly. The diode laser is in the laser pen shown at the *lower left*. Its light is carried up the light path and is focused on the disk. Light reflected from the disk is intercepted by a beam splitter and directed to a quad detector that sits just *above* the laser pen. (*Bottom left and right*) These photographs show a cervix before and immediately after laser vaporization conization. The light areas surrounding the opening in the cervix are the growths that may become cancerous if not removed. After removal, the area is charred from the laser burns.

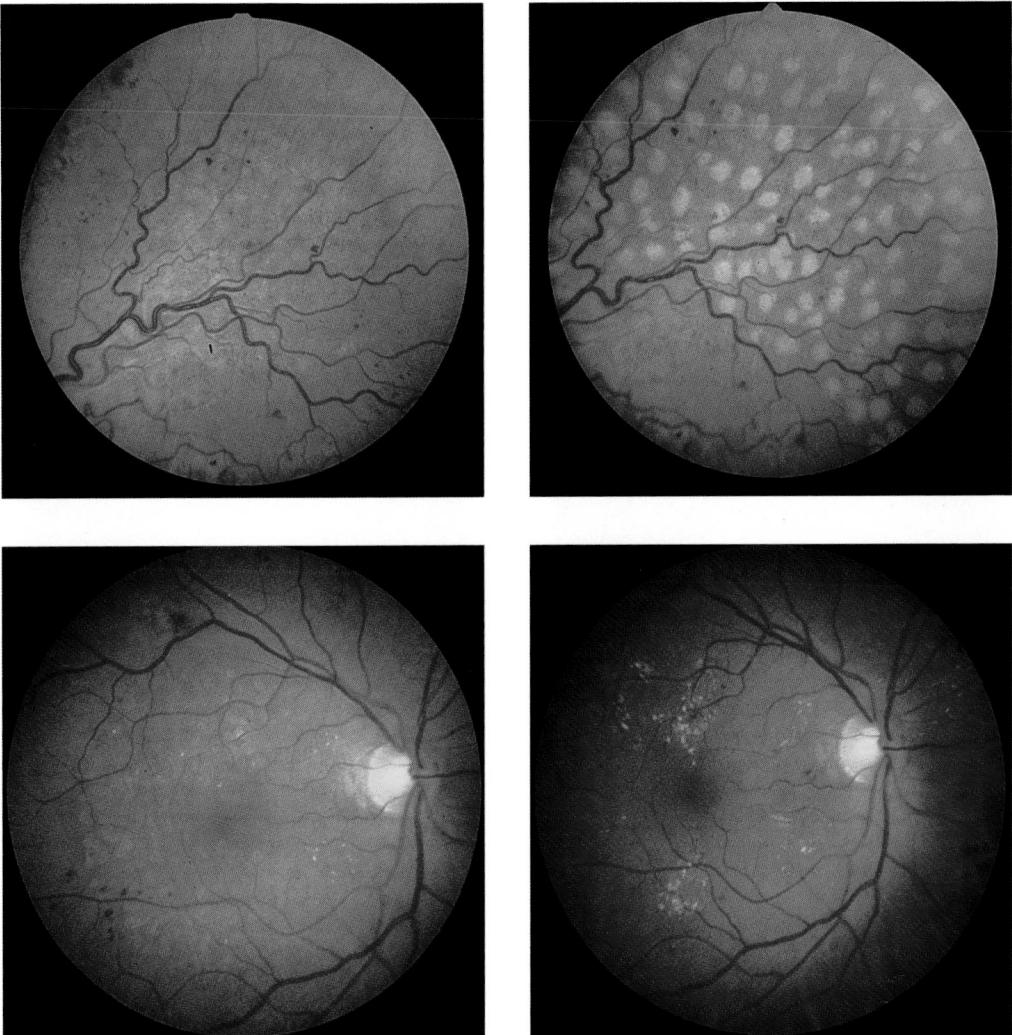

(*Top row*) These photographs illustrate the use of panretinal photocoagulation by an argon laser to treat a form of proliferative diabetic retinopathy. The first photograph, before treatment, shows the region to be treated. In the second photograph, taken just after treatment, the lighter spots are the laser burns. (*Bottom row*) These photographs show argon laser treatment of diabetic maculopathy. The macula is the circular area encompassed by the two most prominent arcs of blood vessels. The central part of the macula, the fovea, is the dark area just to the *left of center* in each photograph. In the first photograph, before treatment, the white exudates are easily visible. In the second photograph, some time after treatment, the exudates have been absorbed. Laser-induced scarring in the region of the fovea can just be seen. The laser spots are very small and slightly lighter in color than the retinal background.

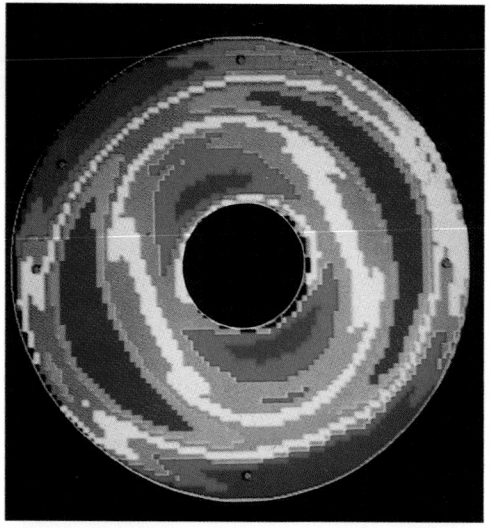
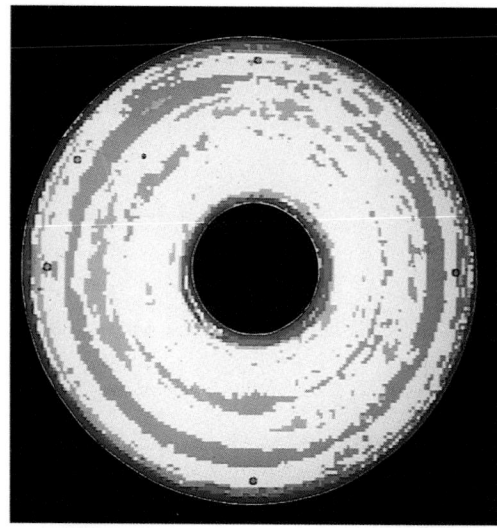

(*Top left and right*) These topographic maps are derived by computer processing of laser interferograms of the space telescope mirror surface. The first map, taken before polishing, shows the high points in blue and the low points in red. The second map shows the mirror surface after polishing has been completed. (*Bottom left*) This colorful figure is formed with scanning mirrors using argon and krypton lasers. The figure develops, grows, moves, and changes colors, and the evolution of the figure can be synchronized to music. The control is by means of taped analog signals. (*Bottom right*) Laser beams and fireworks set up a spectacular display of colors at Stone Mountain Amusement Park in Atlanta, Georgia.

the crystal. The term *solid-state laser* is not meant to include semiconductor lasers which, even though solid, operate on entirely different principles. Solid-state lasers have several advantages over gas lasers. They permit a higher density of lasing atoms. With more atoms, the lasing material can develop more gain and produce more power output per unit volume. Gases at high density are very difficult to excite with electric discharges. Solid-state lasers are usually pumped optically, which is not a problem with the denser lasing material. Since solid-state lasers do not require glass tubes under vacuum or supplies of gas for replenishment, they tend to be simpler to build and maintain. The neodymium laser can be operated in either a pulsed or a continuous mode. Since the solid-state laser crystal tends to be able to store more energy, pulsed operation results in considerably higher power levels. The solid-state lasers are easier and more effective to scale to larger sizes. The largest laser in the world is a neodymium in glass (Nd:glass) laser. The first laser ever operated, the pulsed ruby laser of Maiman, was a solid-state laser consisting of chromium ions in an aluminum oxide crystal. This is still a commonly used laser and has the advantage of a visible red output. However, neodymium lasers are even more widely used.

The neodymium laser consists of neodymium ions (ions are simply atoms with a few electrons missing) embedded in a host material. There are many hosts that have been used for neodymium but the most common are yttrium aluminum garnet (YAG) and glass. The names of the lasers are abbreviated as Nd:YAG and Nd:glass. The neodymium is added as an impurity when the YAG crystal or glass is formed. An important requirement for the host material is to be transparent at the lasing wavelength and at the optical pumping wavelengths. It must be able to withstand the high temperatures induced by the flash lamps without distorting, cracking, or changing optical properties.

The most popular neodymium laser is Nd:YAG. It is used in most laboratories and in many industrial and medical applications. The use of the YAG crystal, with its thermal stability, permits continuous operation. Nd:YAG is limited in size and power because of the limitations in size and the optical quality of the YAG crystals that can be grown. A typical Nd:YAG laser rod is a narrow cylinder about 10 centimeters long and about 1 centimeter in diameter. Nd:YAG lasers can provide continuous output powers from a few milliwatts to 100 watts or more, and pulse energies of 100 joules in 20 nanoseconds, resulting in power levels of 5,000 megawatts per pulse. The continuous powers are powerful enough to be used for certain types of materials processing. They can also be built to very small scale. Power levels high enough to be used by the military as laser designators can be achieved by lasers only a few centimeters long. Laser designators are used to illuminate incoming targets for optically controlled missiles or warheads.

Nd:glass is usually used for larger, more powerful pulsed lasers and can be used to do research on fusion initiation. The most powerful laser in the world, the Nova Laser at Lawrence Livermore Laboratories, is able to develop 10 kilojoules per

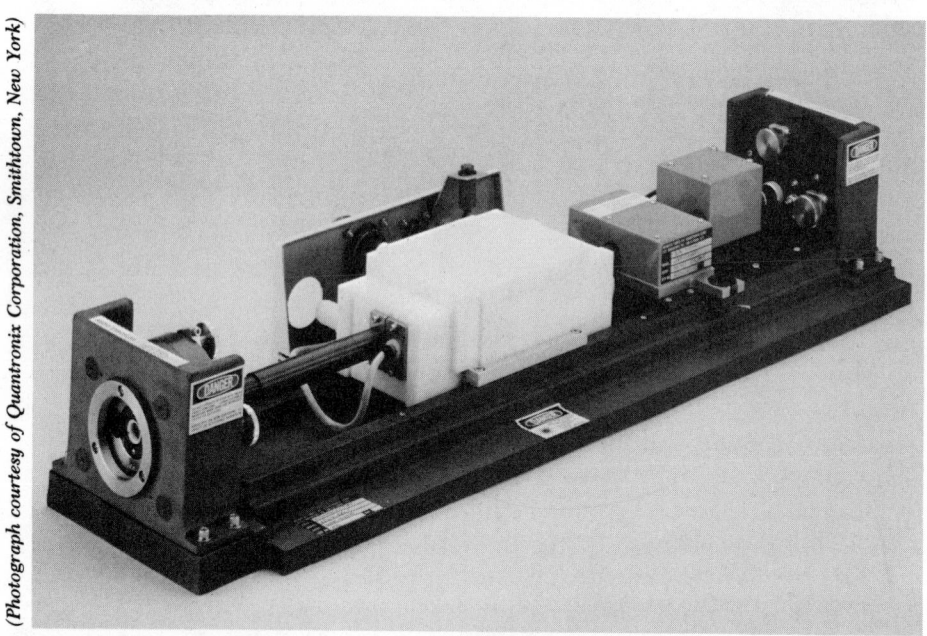

(Photograph courtesy of Quantronix Corporation, Smithtown, New York)

In this Nd:YAG laser, the water-cooled laser rod and flash lamps are contained in the assembly near the *center* of the optical bench. Resonator mirrors, with alignment mountings, are located at each end. To the *rear* of the optical bench are an acousto-optic Q-switch and an aperture-limiting mode selector.

pulse in each of its ten laser arms to concentrate many terawatts of power on small deuterium pellets to start thermonuclear reactions; deuterium is similar to hydrogen and is in the form of pellets because it is frozen. This application of lasers will be discussed in chapter 7. The glass can be fabricated in very large sizes. These lasers usually use oscillators to start a pulse of light and then a series of amplifiers to build it up to very high power levels. Nd:glass lasers are almost always used in pulsed mode, since the glass degrades if it gets too hot. Glass cannot produce average powers as high as YAG and cannot be used for continuous operation. Smaller Nd:glass lasers can produce many pulses per second, but the large ones, such as fusion lasers, can sometimes be fired only once an hour or so.

Almost all neodymium lasers are pumped optically by lamps. Electrical discharges cannot be used with solid-state crystals. The continuous lasers are pumped by tungsten lamps or arc lamps. These lamps convert the electrical energy applied to them into a very intense visible optical emission. Pulsed neodymium lasers are usually pumped by pulsed flash lamps that are excited by power supplies. This provides for higher-pulsed output power levels. The optical emission from the pump lamps is very broadband in wavelength, and the photons are absorbed into two wide energy levels, labeled as levels 3 and 4 in figure 6.4.

Levels 3 and 4 quickly decay in a nonradiative manner into level 2. Level 2 becomes population inverted with respect to level 1 and creates plenty of gain for laser emission. The output is at 1,064 nanometers in the near infrared region of the spectrum. Neodymium can be made to lase at other wavelengths, but they are much weaker; almost all neodymium lasers operate at 1,064 nanometers. This is 1/10 the wavelength of CO_2 lasers, but it is still in the infrared region of the spectrum and not visible to the human eye without an infrared imager. Level 1 quickly decays in a nonradiative manner to the ground state at level 0. Neodymium requires no support from other types of atoms to sustain laser action. The crystals with neodymium embedded in them must be grown with great care to assure purity and uniformity, otherwise, the laser power is degraded and the crystals will decay due to the

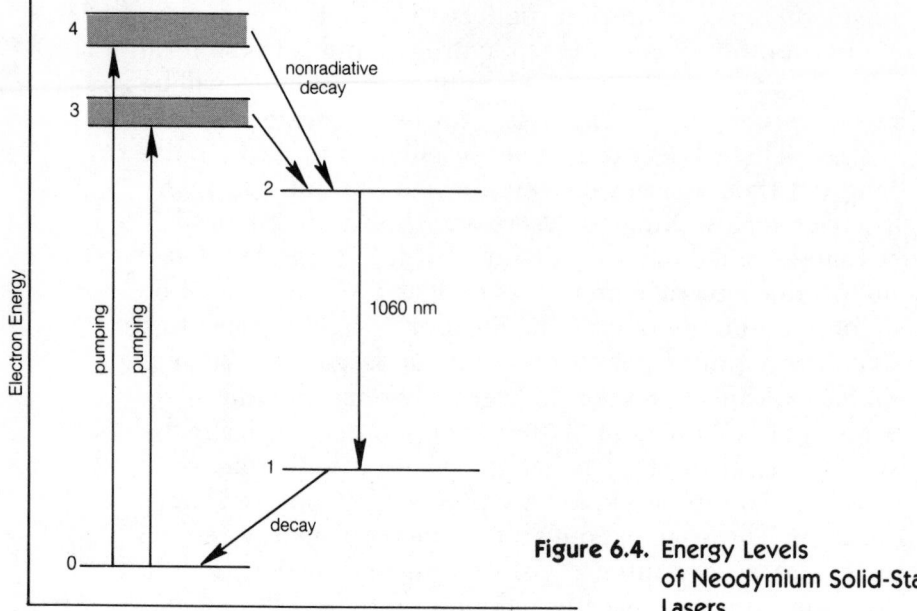

Figure 6.4. Energy Levels of Neodymium Solid-State Lasers

intense heat from the flash lamps. In continuous and high-repetition-rate lasers, water is usually used as a coolant of the laser rod and flash lamps.

A novel and interesting capability of neodymium lasers is that they can be pumped by focused sunlight, which could be very useful for certain applications, especially for laser operation in space. This capacity eliminates the need for any sort of electrical power supply to run the laser, and in space the sun can be available twenty-four hours a day. Space systems could be designed with smaller, simpler power supplies because the lasers could operate on their own. Such lasers might be used for communications between satellites, space stations, and spacecraft, for locating and positioning objects in space, and perhaps for destroying objects in space. These applications, however, would require very large optical systems to concentrate enough energy in the crystals to drive high-power lasers. A more reliable means is to use a diode laser array for pumping the solid-state laser.

For applications requiring shorter-wavelength output, nonlinear optical crystals can be used in conjunction with neodymium lasers to frequency-double the output to 532 nanometers and frequency-triple the output to 355 nanometers. Although some energy is lost in the process of conversion, the higher the power levels are, the more efficient is the conversion process. For very-high-power pulsed lasers, approximately 80 percent of the 1,064-nanometer energy can be converted to 532-nanometer energy in green output. As we shall see in chapter 7, the green output is more efficient for coupling into the fusion targets.

Neodymium lasers are more efficient than HeNe lasers but considerably less efficient than CO_2 lasers, converting about 0.1 to 1 percent of the input energy into output optical power. There are many electrical and optical conversion inefficiencies in the power supplies, flash lamps, and crystal rods that limit the conversion efficiency of the lasers.

Neodymium lasers can be very effectively Q-switched and mode locked. The rods momentarily store up pump energy and build up population inversion. When the resonator is restored, the excited electrons quickly dump their energy, resulting in higher power pulses. Ordinary pulsed operation from a pulsed lamp produces peak powers in the 10- to 100-kilowatt range with pulses lasting about 1 millisecond; whereas with Q-switching, the pulses are many hundreds of kilowatts and last less than 20 nanoseconds. Mode locking produces trains of very short pulses that last less than 200 picoseconds. Because Nd:glass lasers have a broader linewidth, they can be mode locked to produce pulses lasting less than a few picoseconds.

The linewidth of a commercial Nd:YAG continuous laser is about 30 to 150 gigahertz, corresponding to about 10^{-5} nanometers in wavelength. Various devices can be added to the laser resonator to narrow the linewidth. It is now possible to get the linewidth down to about 600 megahertz, corresponding to about 10^{-7} nanometers in wavelength.

The output of neodymium lasers, like that of HeNe lasers, can also be highly collimated. The beams may diverge in a range from about 0.5 degree to less than .01 degree.

Neodymium laser systems can last many years, but their pump lamps must be replaced regularly. On the average, continuous lamps last for only a few hundred hours of operation.

Many commercial neodymium lasers are available, ranging from small systems costing a few thousand dollars to large, metal-working industrial lasers costing hundreds of thousands of dollars. These lasers and other specially designed lasers are used for spectroscopy, development of new laser sources, fusion initiation, and materials development. The military uses them for ranging, designation, and communications. Industry uses them for the cutting, drilling, forming, and welding of plastics, cloths, metals, and other materials. Several applications of neodymium lasers are explained in detail in chapter 7.

Dye Lasers

The lasing medium in a dye laser is a fluorescent organic compound called a *dye* in a liquid solvent. The most important aspect of dye lasers is their wavelength tunability. This means that the wavelength of their output can be scanned over 10- to 40-nanometer ranges. Using different types of dyes, laser output can be obtained at wavelengths ranging from the near-ultraviolet to the near-infrared. The tunability of the dye laser arises from the fact that the dye is a complex molecule that has vibrational and rotational energy levels as well as electronic energy levels. An energy level diagram representing a typical dye is shown in figure 5.1–c. The electronic levels are represented by transitions between the groups marked 0, 1, 2, and 3. The vibrational levels are superimposed on the electronic levels to form bands of levels. The bands are really discrete individual levels, but there are so many of them and they are so close together that they look like a continuous band surrounding the electronic levels. Emission from the dye medium is tunable because transitions can occur from almost any point in the band of level 1 to almost any point in the band of level

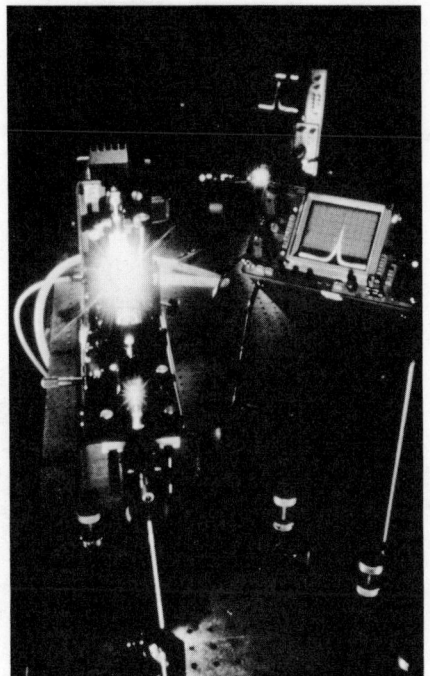

A dye laser pumped by another laser produces visible light in the yellow region of the spectrum. A scientist is tuning an experimental dye laser. The pumping energy from another laser is entering from the *lower left*. The yellow dye-laser beam originates at the bright spot at the *lower left of center* and is reflected by mirrors at the *upper right*. Just *below* is a tube that circulates the dye through the dye-laser resonator.

0. The actual output wavelength is determined by the resonator Q. Prisms or gratings in the resonator will favor high Q at only one particular wavelength at a time. Rotating the prism or grating tunes the laser to the desired wavelength.

Since dyes are liquids, they cannot sustain an electronic discharge and therefore cannot be pumped electrically. They must be pumped optically. Establishing population inversion in level 1 requires pumping to levels 2 or 3. Thus, photons of higher energy than those that are going to be emitted by the laser are required. Higher-energy photons, as we saw in chapter 4, come from short-wavelength sources. Thus, the dye laser has to be pumped by optical sources in the green, blue, or near-ultraviolet regions of the spectrum. Another problem with pumping dye lasers is that the molecules do not easily absorb the light. Scientists say that they have *low-absorption cross-sections*. This simply means that the light tends to pass through without affecting the dye. To compensate for this, the pump light must be of high power and be very concentrated.

The term *intensity* is used to describe the concentration of light. When laser light is focused by a lens, the density of energy in the beam is greatly increased. High- or low-energy-density light is referred to as *high-intensity* or *low-intensity* light. Flash lamps can be used to pump dye lasers and commercial flash-lamp-pumped dye lasers are readily available. But, after what we have learned about lasers, what could better pump a dye laser than another laser? The very first dye laser was pumped with an argon-ion gas laser that emits in the blue or green region of the spectrum. Nitrogen lasers with high-power pulsed output in the ultraviolet region have also been used for this purpose. The most practical lasers now used to pump dyes are either the excimer laser, which we will discuss next, or the frequency-doubled Nd:YAG laser, which we talked about earlier in this chapter.

With laser-conversion efficiencies so low, it does not seem practical to pump one laser with another. However, a typical dye laser can convert up to 50 percent of the input laser-pump energy into output energy. The real value of the dye laser is in its tunability which, when needed, just cannot be obtained in a more practical way. The dye laser is used mostly for research in spectroscopy, where tunability is much more important than output power.

Dye lasers can be operated in three different ways. First, they can be operated for continuous output with relatively narrow linewidth, down to less than 0.001 nanometer, which is a few megahertz. Continuous-output power of dye lasers is low, usually less than 1 watt. Continuous dye lasers can be pumped by lamps, argon-ion lasers, or frequency-doubled neodymium lasers. The second way to operate dye lasers is to pump with pulsed sources such as flash lamps or frequency-doubled, Q-switched neodynium, nitrogen, or excimer lasers. Pulsed-output dye lasers using flash-lamp pumps have energies up to 1 joule, time durations in the microsecond region, and powers up to nearly a megawatt. With Nd:YAG pump lasers, the pulse energies are 0.1 joule and with excimer pumps they are about 0.01 joule. These pulses have time durations of a few nanoseconds. The third way to operate the dye laser is to mode lock it. The pump laser can be continuous, but best results are obtained when the pump laser is also mode locked and synchronized with the dye laser. Typical mode-locked dye-laser pulses are a few picoseconds in duration, with energies of about 10 nanojoules. This corresponds to pulse powers of 10 megawatts. Scientists have built special types of dye lasers called *colliding-pulse mode-locked lasers* whose output pulses are the shortest time-duration phenomena ever seen. They last only 27 femtoseconds. (A femtosecond is a millionth of a billionth of a second.) This laser has been used to measure a transient state of benzine gas with a lifetime of 70 femtoseconds. This is the shortest-duration phenomenon ever measured.

Dye lasers are limited in their average output power not only by their poor absorption characteristics but by the fact that the molecules will disassociate if they are hit by overly intense pump lasers or flash lamps. Because of this and the necessity of cooling the dye, most dye lasers circulate fresh dye in the laser resonator during operation.

Dye lasers are relatively costly and complex. Their primary importance is for research in a variety of special areas in physics, chemistry, and medicine. New applications are opening up in industry and medicine. Dye lasers are used for research in spectroscopy and for special types of measurements. Spectroscopy involves the detailed examination of the structure, energy states, interactions, vibrations, and rotations of atoms and molecules. Dye lasers are used to measure how long atoms or molecules stay in excited energy levels and the rates at which these levels decay to other levels. This type of research is leading toward the discovery of new candidate substances for laser mediums for the building of new types of lasers. The ultrashort pulses of mode-locked dye lasers are used to measure properties of semiconductors that are used to form integrated circuits for computers. They are also used to study rates of chemical reactions as in photosynthesis and biomolecular processes. Medical applications of dye lasers include studies of color vision, ophthalmology, dermatology, cancer treatment, and individual cell physiology.

Excimer Lasers

The term *excimer laser* refers to a type of gas laser that employs an unusual lasing phenomenon. The lasing medium is a molecule that can be formed from excited atoms of two types of gases, one called *rare gases* and the other called *halides.* These rare-gas halide lasers are important because they emit powerful pulses of light whose wavelengths are mostly in the ultraviolet region of the spectrum. Not only are the pulses very powerful, but they can be focused to very high power densities. Furthermore, the ultraviolet photons possess more energy for interaction with atoms and molecules, and for excitation of nonlinear processes generating light output at other wavelengths.

The rare gases are gases like argon, krypton, and xenon, while halides are gases like fluorine, chlorine, and bromine. The rare gases can combine with the halides to form molecules such as argon fluoride, krypton fluoride, and xenon chloride. The interesting point is that these molecules do not exist in nature. They do not form unless the atoms that combine to form them are excited; and the molecules themselves exist only while they are excited, for a few billionths of a second. As soon as the molecules lose their energy, they disassociate back into their constituent atoms. As the molecule collapses, it emits pulses of light. The excited diatomic (meaning a molecule with two atoms) molecules are called *excited dimers*, a term that has been shortened to *excimer.* When the molecule drops to the ground state, the two atoms that were attracted to each other begin to repel each other, and the molecule breaks up. This greatly benefits the laser by keeping the ground state unpopulated. This means that every excited molecule contributes to the population inversion. There is no absorption of laser light by molecules in the ground state as could occur in other types of lasers. In this laser, there is only spontaneous and stimulated emission, no absorption. These factors lead to the development of very high gains in the laser resonator. This type of laser can have sufficient gain to almost achieve laser action without a resonator.

To make the laser work, a rare gas and a halide are mixed

with an inert gas, such as helium, at high pressure, up to five atmospheres. The gas mixture can be excited by either a transverse discharge or an electron beam. The transverse discharge is similar to that used with the TEA CO_2 laser that we talked about earlier in this chapter. As a matter of fact, many lasers designed for TEA CO_2 can be used as excimer lasers, although the excimer-type gases are corrosive and usually require special materials for the gas tubes, electrodes, and resonator mirrors. The electron beam, or e-beam, is the best way to pump an excimer laser; however, e-beam devices are very large, expensive, and impractical for many applications. The e-beam consists of a stream of electrons which must be formed in a vacuum. Electrons are gathered from a cold-cathode field-emission diode capable of delivering e-beam pulses of very-high-energy electrons. The electrons then travel through a thin metal membrane, which serves as the anode, into the high-pressure gas mixture of the lasing medium. E-beam-pumped excimer lasers are more efficient than discharge-pumped lasers and can produce higher output pulse energies in the laser beam. There are many commercial excimer lasers available using the transverse-discharge method of excitation, while e-beam-pumped lasers are primarily found in research and development laboratories. All excimer lasers operate as pulsed devices.

Since excimer lasers have very high gain, they can put out their energy with a very-low-Q resonator. Most excimer lasers use a high-reflection mirror at the back of the resonator with optical coatings at the back of the mirror to prevent exposure to the corrosive gases. The front mirror is usually uncoated, leaving it with a reflectivity of about 4 percent of the laser energy back into the resonator. Because of the high gain, this reflectivity provides sufficient Q for the resonator to oscillate. By contrast, most gas-laser output mirrors feed back 95 percent of the energy into the resonator to achieve laser oscillation. Most commercial excimer lasers produce pulses ranging from 3 to 35 nanoseconds in duration. Research lasers have produced pulses as short as 0.35 nanosecond to as long as 100 nanoseconds. The maximum energy per pulse produced by commercial lasers is about 1 joule. This energy in 16 nanoseconds corresponds to 60 megawatts. The output wavelength of most excimer lasers falls between 150 and 350 nanometers. Some examples of output wavelengths are:

- Argon fluoride at 193 nanometers
- Krypton fluoride at 249 nanometers
- Xenon chloride at 308 nanometers
- Xenon fluoride at 350 nanometers

One type of excimer laser based on atomic fluorine has its output in the near infrared region. Excimer lasers can convert

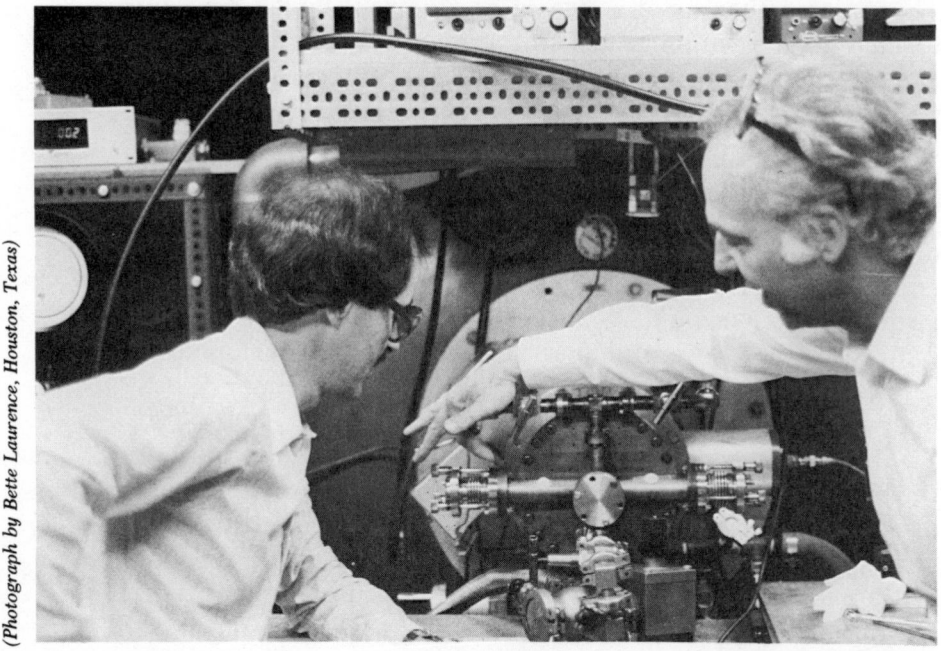

(Photograph by Bette Laurence, Houston, Texas)

Professor Frank Tittel (*right*), of the Rice University department of electrical engineering's Quantum Electronics Laboratory, explains to the author the design of a high-power electron-beam-pumped rare-gas halide excimer laser. The actual laser device is the horizontal tube approximately 1 foot long and 2 inches in diameter. At each end of the laser gas chamber, bellows tubes house the laser mirrors. These tubes permit alignment of the mirrors. High-energy electrons are injected into the laser chamber from a 1-million-volt electron-beam accelerator. The accelerator is the huge steel chamber *behind* the laser equipment. When used to pump a xenon fluoride laser, this apparatus produces output energies of several joules per liter of gas in pulses lasting 10 nanoseconds.

from 1.5 percent to 2 percent of the net energy that goes into the laser into output light energy. Thus, they are more efficient than low-pressure gas lasers like the helium-neon. Because the molecule disassociates when the ground state is reached, the energy levels in the ground state are unpredictable; therefore, excimer lasers have relatively wide linewidths of about 0.3 nanometer. Their output wavelengths are tunable in a range of 3 nanometers by changing the resonator Q. The beam divergence of excimer lasers ranges from 0.1 to 0.2 degree, making them less collimated than many lasers.

Although commercial excimer lasers are available for as little as $15,000, most are very expensive. Hence, most applications for these lasers are in research, with a great potential for use in industry and medicine. These types of lasers will become more attractive for uses outside of research when the technology for building them is better developed. Because of the short wavelengths they are capable of producing and the fact that they can be tightly focused for high peak-power intensities, excimer lasers are attractive candidates for materials processing, for semiconductor etching and annealing, for surface treatment of metals, and for remote sensing. In addition, as a result of the very clean way the high-energy photons can decompose biological tissues without burning surrounding tissues, excimer lasers have significant potential for surgical applications in medicine.

Research applications for excimer lasers include photochemistry, spectroscopy, and remote sensing. In photochemistry the lasers are used to investigate the structures and energy distributions of molecules that are formed in the intermediate stages of chemical reactions. Since these molecules may exist for only very short time periods, they are very difficult to study. In remote sensing, the powerful ultraviolet pulses are directed into the atmosphere and optical sensors are used to pick up light scattered and reemitted from molecules the laser light excites. The reemitted light is called *fluorescence.* Analysis of this scattering and fluorescence from atmospheric molecules can reveal the presence of pollutants and can measure their concentration.

Semiconductor Lasers

The use of semiconductors to form solid-state devices replacing vacuum tubes has enormously reduced the size of a computer. Early computers had to be housed in large buildings. Now you can carry computers or use them on your desk. The semiconductor diode laser is to the gas laser what the transistor is to the vacuum tube. Although the diode laser has not replaced other lasers as the transistor has replaced vacuum tubes, it has its own special uses. As development continues it will tend to replace other lasers when it can do their jobs, because it is almost always smaller, cheaper, and more easily powered. Almost all diode lasers have output wavelengths in the infrared region of the spectrum. Visible-output diode lasers are a recent development. The Japanese have developed room-temperature gallium aluminum arsenide (GaAlAs) diode lasers that operate at 633 nanometers and have output powers from 5 to 50 milliwatts.

Diode lasers lack the beam quality of most other lasers. Their divergence is much larger than gas or solid-state lasers; they easily operate on higher-order transverse modes, resulting in poor output-beam shape; they easily switch axial modes; and they vary in output wavelength depending on temperature. Their advantages, however, far outweigh their disadvantages and make possible many new applications. Diode lasers are used primarily for fiber-optic communications, which we will discuss in chapter 7. They are also used for the recording, writing, and reading of information in the form of laser printers, video disks, and audio disks. These lasers are used in many military systems, such as in battlefield simulation systems where the output light simulates the firing of a weapon. As visible-wavelength diode lasers are developed, they will most likely begin to replace helium-neon lasers in many applications. The real significance of diode lasers is as follows:

- They can be made very small (in some cases smaller than a grain of salt).
- They are very simple, easy to set up, easy to power, and

easy to design into a system; as a result, they are inexpensive to use.

- They are reliable and can be used for applications requiring long lifetimes.
- They can easily be modulated by varying their power supply, unlike other lasers that need special modulation devices.
- They are very efficient in their conversion of electrical power into light output.

After diode lasers were invented in 1962, ten years of research and development were required to produce versions that could be operated reliably at room temperature. The first diode lasers operated at liquid-nitrogen temperatures, and their output lasted only a few seconds. Finding the correct materials, learning how to make the different types of junctions, and developing processes to properly form and deposit the materials for a diode laser was a long and very complex process. A complete understanding of the physics and technology of modern diode lasers is a complex undertaking. Here it is possible only to acquaint you with the output capability of the lasers and their uses.

There are two basic types of diode lasers: the gallium arsenide (GaAs) type that emits in the near-infrared region of the spectrum, and a lead-salt type that emits in the far-infrared. Here we shall describe only the GaAs diode laser.

GALLIUM ARSENIDE DIODE LASERS

Output powers of diode lasers range from fractions of microwatts to 250 milliwatts continuously. Some special laboratory diode lasers have reached nearly 600 milliwatts. Peak powers of pulsed diode lasers range from 0.5 to 1,000 watts. Diode lasers can be combined as *phased arrays* to achieve even higher output powers, although there is the drawback of poorer beam quality than from individual lasers. Diode lasers that generate outputs of a few milliwatts are usually 1 to 3 percent efficient, while other devices, such as gallium aluminum arsenide (GaAlAs) lasers producing 10 to 20 milliwatts, are up to 10 percent efficient. Higher-power diode lasers are more efficient, because there is a threshold current or a minimum current that must be provided before lasing is possible.

Commercial single-heterojunction diode lasers produce pulses with peak powers of from 1 to 30 watts lasting less than a microsecond. Power levels of up to 100 watts can be obtained with these lasers when they are combined into phased arrays. Commercial double-heterojunction lasers produce continuous powers up to 20 or 30 milliwatts, and more powerful versions are being developed.

Precise control of the output wavelength of a given type of diode laser is difficult. Composition of the semiconductor materials can cause the wavelength of the same type of diode laser to vary by as much as 60 nanometers.

The linewidths of commercial pulsed single-heterojunction lasers range from 2 to 15 nanometers, while those of commercial continuous double-heterojunction lasers range from 0.01 to 10 nanometers. In these lasers the linewidth is wider at longer wavelengths. Under carefully controlled laboratory conditions, diode lasers have been operated with linewidths of 0.00001 nanometer or about 10 megahertz.

Diode lasers are not as stable in frequency as other types of lasers. The output wavelengths are sensitive to temperature. Since they can operate in several different axial modes, they tend to jump from one mode to another as the temperature

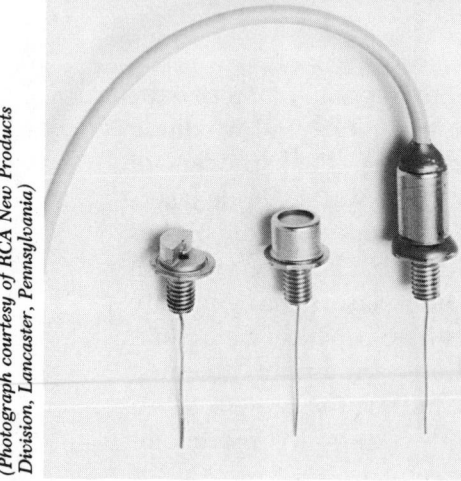

(Photograph courtesy of RCA New Products Division, Lancaster, Pennsylvania)

Three types of diode lasers designed specifically for fiber-optic test systems produce up to 500 milliwatts in 1-microsecond pulses. These are indium gallium arsenide phosphide (InGaAsP) diode lasers that operate at 1,300 nanometers in the infrared region of the spectrum. The laser on the *left* is open, ready for mounting; the one in the *middle* is hermetically sealed; and the one on the *right* is coupled into a fiber-optic cable.

changes, which results in noise in the form of spurious optical signals in communications systems.

Double-heterojunction lasers are important in the new fiber-optic communications systems. Their primary usefulness lies in the fact that they can be modulated at high rates; the higher the rate of modulation, the more information a fiber-optic line can carry. Laboratory devices have been developed that can be modulated at 20 gigahertz. This means that the laser is turned on and off 20 billion times per second. That is, it can be switched in less than 0.05 nanosecond. A single fiber-optic communications system using such a laser could carry a million or more phone calls simultaneously. Typical systems in use today modulate their lasers from a few megahertz to 1 or 2 gigahertz.

The resonator of a diode laser is very short compared to resonators in other lasers, and the diode laser emits from a narrow rectangular area, producing a beam that is oval in shape and rapidly diverging. The beam can spread by a 10-degree angle in the plane of the junction (vertically from top to bottom in figure 5.8) and by a 40-degree angle perpendicular to the junction (side to side in figure 5.8).

Q-switching and mode locking can be achieved with diode lasers, but they are currently being done only in research and development work.

The coherence length for diode lasers is usually short, on the order of a few millimeters. This length is a desirable feature for many applications, but for applications requiring long coherence lengths, like holography (see chapter 7), diode lasers are not a good choice.

Diode lasers require very little input power compared to other lasers. Typically they need only a couple of volts and from 15 to 100 milliamperes of current, which corresponds to less than 1 watt of electricity. This requirement is quite a contrast to that of gas lasers, which often must use many kilovolts to break down the gases and many kilowatts of power to maintain the discharges.

A great deal of research has gone into improving diode laser longevity, an especially important factor if fiber optics are to be used for long-distance communications. Since signals in fiber-optic lines decrease in strength over long distances and must be repeated, diode lasers are necessary to repeat the signals after they have lost their intensity. Repeater laser diodes in transoceanic cable systems are buried at the bottom of the ocean and will have to have very long lifetimes to be cost effective. Double-heterojunction lasers made of GaAlAs emitting a few milliwatts can last from 10,000 to 100,000 hours, or more than 10 years. As new techniques are developed, these lifetimes will increase. There is a trade-off between lifetimes and power levels. The higher the power level of operation, the shorter the expected lifetime. For fiber-optic communications, both high power and long diode lifetime are needed.

Mass-produced double-heterojunction diode lasers can cost

This semiconductor diode laser is small enough to fit within the eye of an ordinary needle. This is one of the types of lasers developed at Bell Laboratories for use with fiber-optic cables in optical communications systems.

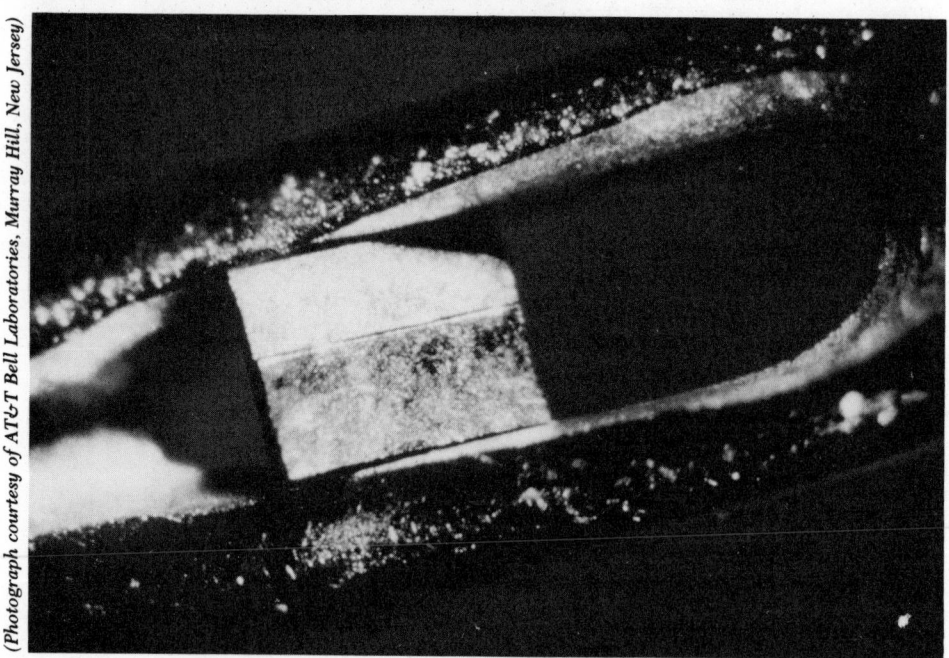

(Photograph courtesy of AT&T Bell Laboratories, Murray Hill, New Jersey)

as little as $10 each when purchased in large quantities, while special high-power or high-modulation-rate lasers can cost a few thousand dollars each.

Since diode lasers are small, are lightweight, and can easily be modulated, their ideal application is in communications. We will talk more about fiber-optic communications in chapter 7. Fiber-optic cables have been developed to the point where they are inexpensive enough to replace old phone lines or to provide new phone lines and computer data links within cities, between cities, or between countries.

Because of their small size and low power consumption, diode lasers are ideal for use in laser video-disk systems, laser audio-disk systems, and computer optical-disk storage systems, in which a relatively powerful laser is used to place information on an optical disk. More information can be placed on one of these disks than can be placed on the magnetic disks currently used with computers. The information can be read back accurately time and time again by a low-power diode laser. For audio enthusiasts, this means having recordings that produce little or no unwanted noise and that can be played almost indefinitely without wearing out. We will talk more about optical information storage in chapter 7.

Other Important Types of Lasers

ARGON-ION LASERS

The argon-ion laser is a gas laser based on electronic transitions in ionized argon gas. Its most powerful output is green visible light at 514.5 nanometers and blue light at 488.0 nanometers. It will lase at several other wavelengths, but at much lower output power. Argon lasers have continuous output and can be mode locked. They are readily available commercially and are used in many laboratories to pump dye lasers. They are used in laser reprographics such as printers and photoresist writing. In law enforcement they are used to illuminate dyes for identifying weak fingerprints. The maximum practical power output for an argon laser is about 25 watts continuously. Most argon lasers produce less than a watt. The argon laser is more powerful than the helium-neon laser, but it is more complex, requires gas replacement, and is considerably more expensive. If the argon is replaced by krypton, the laser emits in the red region of the spectrum. With multiple outputs in the red, green, and blue regions, the argon-krypton laser is attractive for applications in laser art and entertainment. Argon lasers cost from $5,000 to $35,000.

HELIUM-CADMIUM (HECD) LASERS

The helium-cadmium laser is another source of continuous output in the blue region of the spectrum. Its output wave length is at 442 nanometers and is based on population inversion in vaporized cadmium metal. The electric discharge serves to provide pump energy and to keep the metal vaporized. Output powers are considerably less than with argon. Commercial versions produce from 5 to 50 milliwatts continuously. The HeCd laser competes with the argon laser for applications requiring excitation of photosensitive materials, such as in printers, photoresist writing, printed-circuit board inspection, and dye-laser pumping. Argon and HeCd lasers are also used to make audio disks and video disks. Commercial HeCd lasers cost from $4,000 to $12,000.

COPPER VAPOR LASERS (CVL)

The copper vapor laser, another laser with visible output, can operate at either 510 or 578 nanometers, producing green and yellow. As with HeCd lasers, the discharge serves both to excite the gas and maintain the vaporization of the copper metal. The CVL laser has been available commercially for several years. These lasers are simpler than argon lasers and much more efficient. However, they must be operated pulsed. The output laser energy is typically from 2 to 8 millijoules,

This photograph of an argon-ion laser in operation demonstrates the use of various optical devices. The argon laser is to the *lower left of center*. The discharge can been seen at each end of the long rectangular enclosure. Resonator optics are located at the back of the enclosure (to the *left*) and at the front, where the beam emerges. The beam is being directed by mirrors to an optical system at the *upper right*, where the beam is expanded; it then passes through a crystal and a spatial filter, where it is focused down and recollimated to improve the cross-sectional uniformity of the energy distribution.

and the laser pulse repetition rate varies from 4 to 6 kilohertz, which produces average powers of 10 to 40 watts. The pulses last from 25 to 40 nanoseconds. The linewidth is very narrow, approximately 0.006 nanometer or about 6 to 8 gigahertz.

The copper vapor laser has applications in dye laser pumping, combustion research spectroscopy, underwater research, semiconductor research, law enforcement, and medicine. The 510-nanometer output is transmitted well in seawater and therefore can be used for underwater communications or illumination. It can be used to anneal or harden the surface of semiconductors. The CVL, argon, and HeCd lasers are all used to illuminate dyes to aid in fingerprint detection. In medicine, the 577 nanometer output is ideal for treating several skin ailments.

RUBY LASERS

The ruby laser, the first type of laser to operate, is a crystalline solid-state laser, like the neodymium laser. Its output is visible red light at 694.3 nanometers. It is capable of producing very-high-energy pulses. This laser usually must be water cooled to prevent overheating of the ruby rod. Like the neodymium laser, it is pumped by flash lamps. The ruby laser can readily produce millisecond-long pulses of 50 to 100 joules of energy. Since these pulses can be obtained only at low repetition rates, the average power is fairly low. Ruby lasers can be Q-switched to produce pulses of nanosecond duration.

The primary use for ruby lasers is in holography, because the ruby is one of the few sources of high-energy pulses of visible light. Holography is discussed in chapter 7. Ruby lasers are also used for interferometry, for nondestructive testing, and for measurements on plasmas (hot ionized gases). Commercial ruby lasers are readily available but are not as widely used as neodymium lasers. They cost from $14,000 to $200,000.

VIBRONIC LASERS

Another family of solid-state lasers with tunable output are the *vibronic lasers.* Still in the development stage, these lasers are not yet widely used. They are similar to neodymium and ruby lasers in that they are made up of lasing atoms in a crystalline host. However, in the vibronic laser the host crystal permits vibrational energy states to exist between it and the lasing atoms. These vibrational states add to the electronic energy level of the lasing atom to form a band of energy levels, permitting tunability. A simplified schematic drawing of energy levels for a vibronic laser is given in figure 5.1(d).

The most developed type of vibronic laser is the *alexan-*

drite. Alexandrite consists of chromium atoms in a beryllium aluminum oxide ($BeAl_2O_4$) crystal. The output of an alexandrite laser can be tuned from 701 to 826 nanometers.

Other types of vibronic lasers can cover other wavelength regions. One type emits in the ultraviolet, but most, like alexandrite, emit in the near infrared. Only one, titanium in sapphire, has output in the visible range, and it is still in laboratory development.

Why bother with vibronic lasers since dyes easily cover much greater ranges of wavelengths? The reason is that solid-state lasers can produce much higher energy pulses and higher average powers. The alexandrite laser operates continuously or pulsed, producing 100- to 400-microsecond-long pulses with up to 10 pulses per second. The pulses are 0.2 to 0.8 joules each. Average powers range up to 100 watts. This power output combined with tunability makes the vibronic laser unique and gives it many potential applications. However, for the moment, it is still under development even though a few are available commercially.

NITROGEN LASERS

The nitrogen laser, invented in 1963, emits short pulses at high repetition rates in the ultraviolet range at 337.1 nanometers. The emission comes from an electronic transition in nitrogen (N_2) molecules. The laser is pumped by discharge created by very fast high-voltage pulses. The lower energy level of the laser transition has a long lifetime and therefore tends to fill up with molecules in that state. This severely limits the repetition rate of pulses. The laser has very high gain and will deliver its output without a resonator, although a mirror is usually present in the rear of the laser for maximum output.

The nitrogen laser is used for spectroscopy, dye laser pumping, photochemical research, pollution control (remote sensing), atomic and molecular lifetime measurements, medical research, and biological research. It should be noted that this laser is now pretty much obsolete, having been replaced by the excimer laser.

Summary

Some of the most important characteristics of the lasers we have discussed are summarized in the table below. Since laser technology is one of the most rapidly growing fields today, these characteristics are constantly subject to change. The rate at which new types of lasers are being discovered has tapered off slightly since the 1960s, but new development of applica-

Summary of Laser Characteristics

Laser	Type	Common Wavelengths (nanometers)	Max output Power (W) C-continuous P-pulsed	Maximum Efficiency (percent)	Cost ($)
HeNe	gas	543.5 632.8 1152	0.050 C	0.1	200– 10,000
CO_2	gas	10600 9600	30,000 C (400,000)*	20	2,500– 1,000,000
Nd:YAG	solid-state	1060 530[†] 353[†]	200*10^6 P 500 C	1	3,000– 200,000
Nd:glass	solid-state	1050 525[†] 350[†]	500*10^12 P		3,000– 176,000,000
Dye	liquid	380– 1000	1.0*10^6 P 0.200 C	1 (50)[‡]	3,500– 60,000
Excimer	gas	157, 193, 222, 249, 308, 350,	20*10^9 P* 100 avg.	2	25,000– 100,000
Diode	semi-conductor	633– 30,000	0.600 C 100 P	10	10– 3,000
Argon-ion	gas	488.0 514.5	25 C	.01	5,000– 25,000
HeCd	metal vapor	442.0	0.050	.06	4,000– 12,000

* Experimental versions
† Achieved by nonlinear conversion
‡ Conversion efficiency of input laser energy
  Source: Data partially derived from *Laser Focus* and *Lasers & Applications* magazines.

tions has not. Since the potential of most materials to serve as laser mediums has been investigated, it has become more difficult and time consuming to find new acceptable laser materials. Nevertheless, who knows how many hidden possibilities remain? The technology of existing lasers is being rapidly improved and developments like the free-electron laser release us from dependency on achieving population inversion in atoms or molecules.

The next chapter takes a look at some important selected applications of lasers. Few areas of modern life are untouched by lasers. From research to medicine to industry to the military to entertainment, the laser's impact on our lives is enormous and continues to grow.

7

A Myriad Laser Uses

IN the years since 1960, when the first laser fired its unique and powerful pulses of light, the development of applications for lasers has been astounding. At first the laser was considered a laboratory curiosity, a device looking for a use. Very soon, however, as new and different lasers emerged from research laboratories, it became obvious that the invention of the laser was a milestone in the emergence of modern technology, ranking near the same level of importance as the invention of the transistor. We now wonder if there is anything the laser cannot do. It has influenced almost every other area of technology, even if only indirectly. Besides contributing to advancement in thousands of areas of technology, the laser has given birth to new technologies that would not have been possible without it. These new areas include holography and nonlinear optics, which are discussed in this chapter. One of the most interesting and unexpected areas to be significantly affected by the laser is medicine. Even the most farsighted laser pioneers admit that the use of lasers in medicine was an unexpected development.

In this chapter, we will not be able to talk about all applications of lasers but will choose a few important ones from research, engineering, industry, medicine, communications, business, and the military. New uses for lasers are arising so rapidly that it is necessary to follow technical magazines and publications to keep up with developments.

Applications in Research and Development

LASER SPECTROSCOPY

Spectroscopy is the study of materials by observing the way light interacts with the matter. A beam of light is directed into a gas, liquid, or solid, and very sensitive measurements are made of the spectral content (wavelengths and linewidths), the intensity, and the direction of the light absorbed or emitted by the material. It is through spectroscopy that we measure the characteristics of electronic energy levels in atoms and vibrational-rotational energy levels in molecules.

Many things can happen to light when it interacts with atoms and molecules. It can be absorbed without the occurrence of other light emissions. It can be absorbed, and new emissions can occur due to other energy transitions in the atom or molecule, a process called *fluorescence.* It can be scattered, meaning that it can be diverted into other directions. It can result in the ionizing of the material. The wavelength of the excitation light may be changed when it is scattered. Two photons can excite an energy level with twice the energy-level difference of one photon. Many other effects can occur, but will not be discussed here.

It is in spectroscopy that ultrashort nano- and picosecond pulses from lasers become important. They permit the spectroscopic study of ultrafast chemical reactions and short-lived reaction products such as those involved in vision and photosynthesis. Ultrashort-duration biochemical events in processes like photosynthesis are studied by means of short-pulse duration excitation at carefully chosen wavelengths and the examination of the subsequent emission and absorption.

The various types of spectroscopy are divided according to the type of interaction being observed: absorption spectroscopy, fluorescence (emission) spectroscopy, various types of scattering spectroscopy, Raman spectroscopy, and two-photon spectroscopy. Raman spectroscopy is the study of the scattering that results in the emission at wavelengths different from the excitation wavelength but not due to other transitions.

Two-photon spectroscopy and Raman spectroscopy are based on nonlinear effects. With linear effects, the wavelength of the incident light never directly changes, and only one photon at a time participates in the interaction. In nonlinear interactions, wavelengths can directly change; in effect, photons are split or combined to form new photons of different wavelengths. In other types of nonlinear effects, two or more photons participate simultaneously in the interactions. The two photons may be from different lasers. With nonlinear effects, much-higher-input excitation-light intensities are usually required, and the output emissions from the materials are lower in intensity and more difficult to observe. Linear effects are stronger and even though new emissions from the material may appear, they are not the result of direct wavelength conversion but are due to other energy transitions.

The interaction of light with materials and the resulting emissions are critically dependent on the wavelength and linewidth of the excitation light. The more intense the excitation, the more intense and more easily observable will be the resulting emissions. Ideally, spectroscopy requires an intense input beam at a wavelength required to excite a specific energy-level change (transition) in the material. The beam should have the narrowest possible linewidth to assure high-resolution results. Is it any wonder that the science of spectroscopy was revolutionized by the invention of the laser and that the laser is now the most important source of excitation light for spectroscopic studies? Because the excitation light must be matched to the energy levels being studied, the dye laser, with its tunability, is a very important development. The dye laser is now used primarily for spectroscopy and has greatly expanded the capability of the science. Many more atoms and molecules can now be investigated, with the result that more can be learned about their various energy levels and the way they chemically react to combine or disassociate.

Spectroscopy is the science of uncovering detailed information about the energy levels of atoms and molecules. It is used to determine the value of the energy levels, the separation between levels, how long the electrons stay in the level when they are excited, and what happens as the atom or molecule returns to the ground state. Besides wanting to know how long energy levels last, scientists also are interested in whether or not decaying energy levels emit radiation, what the wavelength, intensity, and linewidth of the emitted radiation is, whether or not the levels are singular or in bands, whether or not the levels are split into sublevels, and how the energy emitted is divided between electronic, vibrational, rotational, and translational forms of energy. All this information reveals a great deal about why materials behave the way they do. It points the way to finding new and valuable optical and electronic uses for materials.

In a sense, spectroscopy is the science of laser materials, and its most valuable tool is the laser. Spectroscopic information on materials is the information scientists have used to discover many new lasers. It has helped them to invent lasers that could output new and different wavelengths, and to do it with greater power and efficiency. Scientists do not try to make lasers out of just any material. In most cases they know they have a pretty good chance of getting the material to lase at one or more wavelengths, and they know how to excite the material. Of course, once laboratory experiments are under way, a great many new and unexpected things are discovered about how materials perform as lasers. It is the knowledge gained from spectroscopy that leads to such rapid development of so many different kinds of lasers with so many different capabilities.

Beyond finding suitable materials with which to build new lasers, spectroscopy performs several other functions. It helps us to determine the composition of the sun and stars, the rate at which ultraviolet light from the sun changes the composition of the atmosphere, the rates at which chemical reactions take place, and what intermediate compounds are formed during the reactions. Chemical reactions can be selectively initiated and studied for possible formation of new compounds. Selectively initiated chemical reactions can be used to separate isotopes, which are very similar forms of the chemical elements that differ in atomic weight (the number of neutrons in the nucleus); they are so similar chemically and physically that they are very difficult to separate. They must be separated, for example, in uranium for use in atomic reactors. Because different isotopes have slightly different sets of energy levels, they can be induced to selectively participate in light-sensitive chemical reactions, after which the isotope is recovered from the reaction products.

Absorption spectroscopy is used to discover and measure the abundance of very minute amounts of impurities in gases. Thus it is a very valuable technique for the monitoring and control of pollution. In one technique, samples of air and water are spectroscopically analyzed for the presence of pollutants. Another technique uses powerful lasers directed into the atmosphere combined with very sensitive detectors on the ground to look for scattering and fluorescence from pollutants; with this technique, called *remote sensing,* samples do not have to be taken, and pollutants can be measured even if they are present to only one part per billion.

Using more powerful lasers, scientists are taking advantage of the nonlinear interactions to learn even more about chemical reactions. This methodology is especially valuable when the reaction environments are harsh and direct sampling is out of the question. Laser beams are being sent into the exhaust plumes of jet and rocket engines and into the combustion chambers of test-automobile engines to measure the temperature of the reactants, to identify reactant compounds, to measure the rates of the reactions and the lifetimes of the compounds, and to identify the different stages that occur in reactions. This information is derived from observing the intensity and frequency variations of the emitted optical signal as two pulsed lasers are used to excite chosen molecular vibrational energy levels. Reactants lasting only picoseconds can be identified. It has never been possible before to so fully investigate the physics and chemistry of molecules in fast reactions such as burning. This kind of knowledge will help us to extract more power and better efficiency from the burning of fuels.

Laser spectroscopy techniques can be applied to precision measurement. They have been used to determine the frequency of visible yellow and red light to within 1.6 parts in 10^{10}, almost 1,000 times better than could be achieved before. Since the velocity of light is known with great accuracy, this means that the wavelength can be determined from frequency measurements using methods much more accurate than with direct wavelength measurement.

Laser spectroscopy is being applied to biological systems. It can be used to detect very small quantities of elements or impurities in biological fluids such as blood. This capability has great potential for medical diagnosis and law enforcement. It is also being used to identify and measure the distribution and movement of molecules in living cells, such as nucleotides, hormones, and cholesterol. Special molecules are created that fluoresce and that are used to replace the molecules usually present. These studies can tell us such things as how molecules move over the surface of cells, how molecules move between cells, and how cells move around. The way hormones and growth factors interact with cell-surface receptors is a very important area of biological research, which may be able to tell us how faulty growth control occurs in cancerous cells.

ENERGY PRODUCTION SOURCES—INERTIAL CONFINEMENT FUSION

As the earth needs more and more energy and we run shorter of fossil fuels such as oil and gas, scientists are constantly looking for alternative sources of energy. Although we have harnessed the energy of the atom bomb—atomic-fission reactions—to build our nuclear reactors, they have proven to be expensive and not sufficiently reliable.

The great prospect exists that we can harness the energy of the hydrogen bomb—the same energy that keeps the sun and stars going, nuclear-fusion reactions—to generate perhaps unlimited amounts of power to satisfy our needs. However, some very difficult technical problems must be overcome in order to do this. Problems arise primarily because isotopes of hydrogen, deuterium, and tritium must be heated to many millions of degrees before the fusion reaction begins. Two methods are currently being researched in an attempt to harness fusion for controlled use. The first is called *magnetic confinement fusion* (MCF), and the second is called *inertial confinement fusion* (ICF). In MCF a plasma of hot ionized gases is formed and held in place by strong magnetic fields, while the gas is heated by powerful radio frequency waves. The devices used for MCF are called *tokamacks*. In ICF, intense laser beams with incredibly short-duration high-energy pulses are focused on small pellets of deuterium and tritium to compress and heat them. At the present time it has not been possible to get significant energy from the compressed pellets, but the feasibility of doing so has been amply demonstrated.

A new laser called Nova that is now being used at the Lawrence Livermore National Laboratory has the potential for getting significant energy back from the compressed pellet, making it feasible to initiate, control, and use fusion energy. The Nova laser is capable of simultaneously firing 10 beams of light onto the pellet targets with a total energy of 100 kilojoules in 3 nanoseconds, which corresponds to 33 terawatts of power; the pulses can be shortened to 100 picoseconds delivering 100 terawatts. When the light pulse hits the pellet, it vaporizes the outer layer so rapidly that the inner portions are compressed to 1,000 times the density of liquid hydrogen. The temperature is suddenly up to many millions of degrees. The fusion reaction begins in the center and proceeds outward. Theoretically it should be possible to completely react the pellet and get the energy out of it before the deuterium and tritium fuel is blown away.

The Nova laser is without a doubt the largest and most powerful laser ever built. It is the largest precision optical project ever undertaken. It contains 1,000 major optical components. The laser is a Nd:glass laser consisting of a master oscillator divided into ten separate beams. Each beam is separately amplified in several stages and filtered before being directed onto the target. At each stage, the beam becomes larger in diameter. The enlargement is necessary because as the beams are amplified, they become so intense that their energy density would otherwise damage the optical materials, coatings, and components; making the beam larger keeps the energy density down. The first-stage amplifiers are single rods of neodymium in phosphate glass that is 4 centimeters in diameter. The last amplifiers are huge elliptical slabs of neodymium glass, two in each amplifier, which are placed at

(Drawing courtesy of Laurence Livermore National Laboratory, Livermore, California)

The largest laser, at least so far, is shown in an artist's cutaway view of the Nova Inertial Confinement Fusion facility. The laser beams begin at the *right* and are divided into 10 separate beams. They pass through various amplifier stages while moving from *right* to *left* in the photograph. As they are amplified, the diameter of the beams increases, as does their energy. At the *right* of the photograph, a system of mirrors directs the beams from different angles into the spherical target chamber. The beams are converted from the infrared region of the spectrum into the green just before entering the chamber.

Brewster's angle to minimize reflection. The slabs are pumped by huge banks of flash lamps that send their energy into the face of the neodymium glass. The aperture of the beam when it passes through these amplifiers is 46 centimeters. There are a total of 16 amplifiers in each of the 10 beams.

Each beam of the laser must also pass through 7 spatial filters, each of which consists of a lens used to focus the beam through a very small aperture and another lens used to recollimate the beam. Spatial filters help to keep the beam uniform across its cross section. Misdirected or unwanted light is filtered out from the main beam. The spatial filters also serve as beam expanders. By using longer focal length lenses on the output of the filter, the beam is allowed to grow larger in diameter before being recollimated. Just before entering the target area, a portion of each beam (about 2 percent) is allowed to escape through a mirror so that the beam can be examined to make sure it is uniform, properly aligned, and of sufficient energy and power. Several relay mirrors then send each beam into the target area from a different direction. The beams are frequency-doubled to 525 nanometers or frequency-tripled to 350 nanometers before being focused onto the pellet targets by fused-silica lenses 74 centimeters in diameter.

The Nova laser has a feature never before incorporated into a fusion laser. It uses huge nonlinear optical crystals of potassium dihydrogen phosphate (KDP) to convert the original laser wavelength of 1,050 nanometers in the infrared region to 525 or 350 nanometers in the green and blue. The efficiency of nonlinear optical conversion is dependent on beam intensity. Since the Nova beams are so powerful, the KDP can convert up to 80 percent of the original energy into frequency-doubled energy (also referred to as the second harmonic) at 525 nanometers. The laser delivers 50 to 80 kilojoules in 3-nanosecond pulses at 525 nanometers, and 40 to 70 kilojoules in 3-nanosecond pulses at 350 nanometers. During experiments with the Nova laser's predecessors, the Shiva and Argus lasers, it was discovered that hot electrons formed that preheated the pellets and prevented full absorption of the laser light, reducing the level of compression that could be achieved. It was discovered that shorter-wavelength light does not produce nearly as many hot electrons and interacts more directly with the hot ionized gases (plasma) being formed from the target. Since the overall absorption efficiency at shorter wavelengths is greatly improved, the Nova laser was designed with frequency conversion.

The optics are only part of the story of the Nova laser. High-energy-density storage capacitors store up to 60 megajoules of energy that are delivered to the more than 5,000 flash lamps that pump the neodymium-in-glass lasing medium. A pulse-synchronization system is required to adjust the optical path lengths to assure that pulses from the ten beams arrive at the target simultaneously within a picosecond. Since maintenance and manual alignment of all these components would be nearly impossible, the Nova laser has an elaborate computer control system consisting of five VAX-11/780 computers that receive alignment data and relay control information to 50 microcomputers that drive electronically controlled (servo-controlled) optical components.

The target pellets of deuterium and tritium are contained in a huge spherical aluminum vacuum chamber 4.6 meters (15 feet) in diameter with walls 13 centimeters thick. The chamber is equipped with many instruments used to measure the return of neutrons and X rays from the pellets. Images are formed to help scientists assess the degree of compression taking place. Other instruments measure various forms of energy release and heating that occur in the deuterium and tritium. With all these measurements, the degree to which fusion is taking place in the experiments can be determined.

As large and powerful as the Nova laser is, it is still not suitable for energy production. It cannot be fired rapidly enough. Its pulse-repetition rate is very low. It is a research tool designed to find out what is required to initiate controlled-fusion reactions. Even though it has high-power pulses, its average power is low. Average power is the amount produced averaged over a long time period. A laser that produces fusion energy will have to be even more powerful and be capable of being fired every few seconds or faster. This produces higher average power. Research efforts continue in the development of lasers for fusion-energy production. Neodymium may not be the answer. Neodymium in glass heats up at high-repetition rates and neodymium in YAG is very expensive. Further, neodymium in YAG crystals would be difficult to grow in such large sizes. Other high average power lasers like CO_2 and chemical lasers have been tried for fusion but have the disadvantage of long wavelength. Short-wavelength high-power lasers are the most likely answer, bringing excimer lasers into focus for development as fusion lasers. Research in high-power excimer lasers is being conducted at Los Alamos National Laboratory. Their krypton fluoride laser, called the Aurora laser, is able to deliver 10.5 kilojoules in a 500-nanosecond pulse. The peak power is 20 gigawatts. Power enhancement methods are being tested to bring the laser power up and to help assess whether or not the excimer laser has a future in inertial confinement fusion.

HOLOGRAPHY

Holography is the science of creating and using three-dimensional images of objects. Holography was invented by Dennis Gabor in 1948, long before the first laser was invented, but without the laser, holography did not go very far. Holography without lasers can be done only on very small or very flat objects. The formation of three-dimensional images of large objects with sizable contours requires light of long coherence lengths, obtainable only with lasers. Holography depends on the coherence and monochromaticity of laser light. Of course, if the light waves are not of very nearly the same wavelength, they cannot be coherent. They will get quickly out of phase; therefore coherence implies monochromaticity. Even though most of the work on holography goes on in research institutions, it has already found some very important applications in industry and is having some limited success in entertainment. In industry, holography is being used for the nondestructive testing of mechanical components through a technique called *holographic interferometry.* This technique makes it possible to measure movements in the surface of an object to within about 0.05 micrometer. In entertainment, the feasibility of generating three-dimensional movies with holography is creating a lot of excitement, although there are many technical problems to be surmounted. The requirements for recording a hologram do not lend themselves well to making movies.

To understand how holograms are made, we must talk about *wave fronts.* When an object is illuminated by an optical source, it reflects the light away from itself, forming a wave front. The wave front is the formation of the optical signal reflected from an object that is being illuminated. Figure 7.1

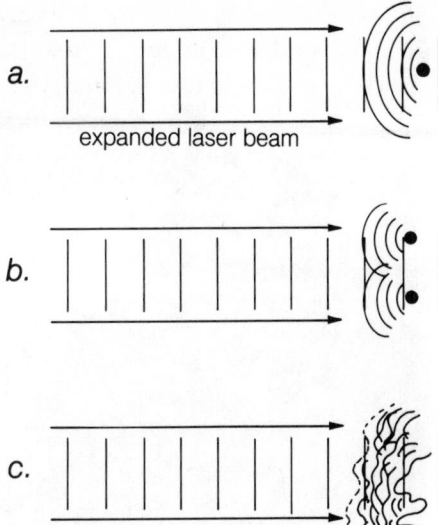

a.

expanded laser beam

b.

Figure 7.1. Reflected Wave Fronts   c.
from Points and
Surfaces of Objects

is a simplified illustration of the illumination of an object and the formation of the wave front. If we use a laser to illuminate the object, we must expand the beam with lenses so that the laser beam will be large enough to flood the object with light. The expanded laser beam forms a plane wave front that is represented in the figure by the vertical lines perpendicular to the direction the beam is traveling. The vertical lines represent lines of constant phase in the laser beam. As we look along the lines of constant phase, we find the waves coming from the laser at approximately the same point in their movements from peaks to troughs. When the laser plane wave front illuminates a point object as shown in figure 7.1–*a*, it is reflected from the point as a returning spherical wave front. In part *b*, if there are two points, they each emit spherical wave fronts and the net result is the sum of the two wave fronts. Optically, any object behaves as the sum of points that make up its surface. As shown in figure 7.1–*c*, an object of arbitrary shape will reflect a wave front that is the sum of spherical waves emitted by each point on the object.

The wave front contains two types of information, both intensity and phase. The intensity will vary depending on the

(Photograph courtesy of Apollo Lasers, Inc., Chatsworth, California)

This holographic camera, used to make holograms and holographic interference fringe patterns, employs a microcomputer-controlled ruby laser and two amplifiers. Laser control parameters are selected in response to user inputs defining the required pulse energy and pulse separation. The ruby laser pulses can be synchronized with the vibrations of an object being analyzed. The pulsed ruby-laser oscillator is the white box at the *upper right.* The long black tube just to the *lower left* of it is an alignment HeNe laser. The two white boxes in the *center* are ruby-laser amplifiers. The reference beam travels back and forth between multiple mirrors in the *center* of the bench to equalize the reference-beam path and object path. The reference beam strikes the film at the *lower right,* behind the notch in the table. The object beam is reflected toward an object located at the *lower left,* off the side of the optical bench.

reflectivity of each point on the object, and the phase will vary depending on the shape of the object. Holograms are capable of recording enough information to fully reproduce the wave front coming from the object. They retain both the intensity and phase information. In an ordinary photograph, the intensity information is retained, but the phase information is lost. When a hologram is played back, it reproduces the original wave front, making it appear that the object is there. You can even move your head around to look at the object from different directions. With a photograph, we have only a two-dimensional representation of the original, which captures the wave front at only one point in space. You cannot gain anything by looking at a photograph from a different angle.

Holograms are recorded through interference that takes place between the wave front from an object and a reference wave front. The reference wave front is the original wave front from the laser, which arrives without striking the object. Among the many arrangements for doing this, one is shown in figure 7.2. As we saw in chapter 4, interference takes place only if we divide up the beam from one source. The laser beam is divided up by a partially reflecting mirror, and then each beam is made divergent by passing it through a lens. One wave front is reflected by a mirror onto the object, while the other is reflected by a mirror directly at the photographic emulsion. The wave front from the object and the reference wave front mix by interference in the photographic emulsion. The emulsion records the resulting interference pattern. This result is a very complex pattern that looks nothing like the original object. As a matter of fact, the interference pattern in the emulsion is the sum of many interference patterns. Each point

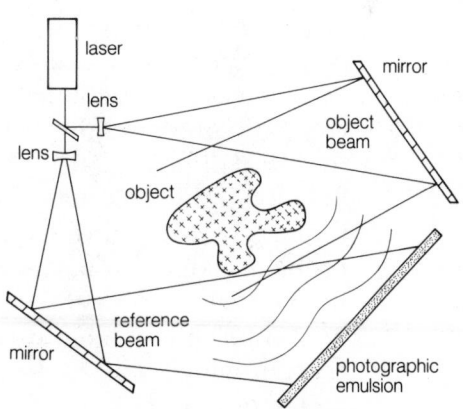

Figure 7.2. Formation of a Hologram

on the object generates a set of waves forming a spherical wave front that produces a pattern of light and dark ridges as the waves add to or subtract from the waves in the reference wave front. Not only is the pattern in the emulsion the sum of all the interference patterns from each point on the object, but each point in the emulsion has a record of interference from each point on the object. When the emulsion is developed, the resulting hologram will possess areas of reflectivity corresponding to the interference pattern of the object and the reference.

Now that we have recorded it, what do we do with it? The answer is really very simple. All you have to do is to develop the film and illuminate it with the expanded laser beam. In doing so, you recreate the original reference beam. When the original reference beam encounters the interference pattern in the hologram, it acts to recreate both of the original wave fronts that originally mixed to form it. Thus if you look through the hologram in the same direction that the object was originally located, you will see a full three-dimensional image of the object, just as if it is present. The reason it looks so different from a photograph is that the complete wave front that originally came from the object is being reproduced. You can move your head around and see features in the object that are above, below, or behind other features. Of course, if you move your head too far, you will see nothing, because there is a limit to the angle with respect to the object that the emulsion was able to record. The object also may not look very real, because it has been recorded in only one color. Indeed the very thing that makes holography possible, monochromaticity, is quite a limitation, since we like to see things in full color.

Let us talk about a few more differences between holography and photography. These differences contribute to the technical difficulties of making three-dimensional movies. Since in holography we can deal with only one color at a time, it will require at least three holograms to create color. A major problem is that if you play back with a green laser the hologram of an object made with a red laser, the image will be a different size. Another problem is that photography is an imaging process, in which we use lenses to image large scenes on small pieces of film. Holography is a very different process; we are mixing wave fronts by interference, and the photographic emulsion has to be as large as or even larger than the object, depending on the size of the angle at which you want to be able to view the image. To form an image that can be played back and viewed from any angle would require a spherical emulsion that would completely surround the object.

Another limitation in recording holograms is vibration or movement. If any of the objects involved in making the hologram move to within less than a wavelength of the light, the hologram will be smeared. Holograms are made with very rigid laboratory optical components isolated from outside vibration by special vibration-isolation tables. Making holograms of moving objects will require very intense short-pulse-duration lasers, which we have; unfortunately, we do not have emulsions sensitive enough to get exposed that fast. A serious problem in making holographic movies is that lasers would be dangerous for the eyes of nearly everyone on the movie set, let alone the actors and actresses. Illumination of a movie set by pulsed lasers alone would probably leave everyone in a state of confusion. In spite of all this, there are many ingenuous ways of getting around these problems, and eventually we will have three-dimensional movies. Already, ways have been found to convert into holograms movies taken while progressing around an object, and cylindrical holograms have been made of objects that permit the observer a 360-degree view.

Now that we know what a hologram is and how sensitive it is to movement of the object being holographed, we can talk about one of the most important uses of holography: holographic interference. Many objects manufactured today are subject to stress and bending. Stress is induced in machines and their components by temperature changes, vibration, and pressure. Manufacturers can improve their designs of mechanical components and uncover defects in their components if they can observe how the components respond to stress. For this purpose, a double hologram is taken, first with the object or component at rest, or unstressed, and then, changing nothing else, with stress applied to the object. It may be heated, pulled on, pushed, twisted, or have anything else done to it to simulate its operating environment. The second hologram is taken in the same emulsion as when the object was unstressed. The result is a hologram with fringes in it called an *interferogram.* The fringes represent movement of the surface that took place from the unstressed to the stressed condition. The movement can be determined to within about one tenth of the wavelength of light used. That is about 0.05 micrometer or roughly 2 millionths of an inch! Obviously, this capability is very useful to industry for perfecting mechanical design. It is being used on aircraft tires, jet engine turbines, automotive transmissions, automotive clutches, computer disk-drive components, high-pressure valves, gears, and automobile frames, to name just a few.

NONLINEAR OPTICS

Nonlinear optics is a field that deals with phenomena that occur only at high light intensities. Before the invention of the laser, the occurrence of these phenomena could only be theorized; they had not been observed. Nonlinear phenomena of the types that take place when intense laser beams interact with optical crystals have analogous effects that are well known to electrical engineers. Circuit elements have nonlinear responses that enable engineers to double the frequency (halve the wavelength) or halve the frequency of a radio or microwave signal and to mix two signals to form the sum or difference frequency of the two. Many of us use a device that depends on nonlinear phenomena at microwave frequencies. High-quality police radar detectors achieve high levels of detector sensitivity by mixing the incoming radar signal with a local oscillator (source of a radar frequency signal) to generate the difference frequency. The difference signal is lower in frequency than the original signal and therefore is easier to use in the circuitry to trigger the warning. The process is called *heterodyning.* The processes are referred to as nonlinear because the frequency (wavelength) of the signal changes; linear processes do not change the signal frequency.

Certain types of optical crystals with nonlinear interaction properties are used to generate the harmonics or subharmonics of laser beams. The harmonics of a beam are beams that have

(Photograph courtesy of Newport Corporation, Fountain Valley, California)

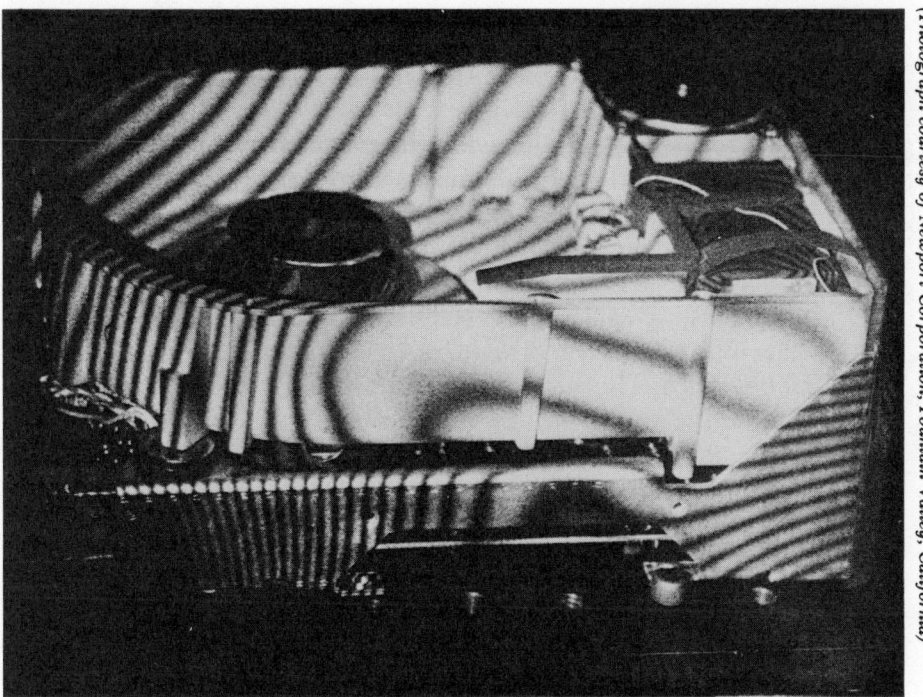

This holographic fringe pattern reveals thermal deformations in a 13 centimeter (5¼ inch) disk drive. Each fringe represents a 0.32-micrometer change in the surface elevation. Holographic contour maps such as these tell engineers how to alter the design of objects to make them less sensitive to changes in temperature.

double, triple, etc., the frequency of the original beam. Subharmonics are beams with one-half, one-third, one-fourth the frequency of the original beam. The crystals can also be used to mix two optical signals to form the sum or difference frequency of the two. Two laws of physics govern the mixing of waves. The first determines the frequency of the wave being formed, and the second governs the way the waves interact in the crystal.

The first law is conservation of energy. Sometimes it is easier to think of nonlinear interactions as the combining of two photons to form a third or the splitting of a photon to form two others. Any time two electromagnetic waves interact to form a third wave, photon energy is conserved. Since the energy of a photon is proportional to its frequency, the sum of the frequencies of two of the waves will equal the third. The formation of the harmonic (frequency-doubling) can be thought of as the mixing of a wave with itself. The frequency of the wave formed is simply twice the original frequency.

The second law is conservation of momentum, which requires that certain conditions be met in a crystal before it can sustain a sufficient level of interaction of the waves to be noticeable. Optical signals of different frequencies (wavelengths) propagate in crystals at different relative velocities. The index of refraction in crystals is different for different frequencies. Therefore, for arbitrary indices of refraction, the interacting waves will be out of phase as soon as they form at a given point in the crystal. Unless they are somehow synchronized, no significant power level can build up in the wave being formed. However, the indices of refraction in crystals depend on the angle between the laser beam and the direction of the crystal axes. Indices of refraction also change with crystal temperature. These last two facts enable scientists to vary the indices of refraction of the crystals so that the beams will maintain their interaction. This contributes to the growth of the wave being formed.

Nonlinear interactions are intensity dependent and become more efficient for higher-power lasers. As has been mentioned, the Nova ICF laser uses potassium dihydrogen phosphate (KDP) crystals to convert up to 80 percent of its incoming infrared beam into the second harmonic. Smaller continuous-output lasers may be able to convert only a few percent of their energy to a different wavelength. High-repetition-rate pulsed lasers are much more efficient for nonlinear optical interactions than continuous lasers, since their pulses are much more powerful.

Nonlinear interactions are enhanced by the use of resonators. The resonator provides feedback for additional amplification of the waves produced by the nonlinear interaction in much the same way that a resonator supports laser action.

Another example of a valuable use of nonlinear interactions is in the development of a laser for submarine communications. The powerful zenon chloride excimer laser developed for this purpose has its output at 308 nanometers. However, this is not the optimum wavelength for the high-discrimination filters and detectors that are designed for use in submarines. The output of the laser is converted to 459 nanometers by a lead-vapor cell. This system is discussed in the section in this chapter on communications.

The use of nonlinear optical interactions increases the number and types of things that lasers can do. Developments in the field of nonlinear optics contribute significantly to advancements in research and to the increased role of lasers in our lives.

Applications in Engineering

Two interesting examples of the value of lasers in engineering are their use for precision measurement and their use for computer systems technology. This section presents the use of lasers for precisely contouring objects, for measuring the motion of aircraft and spacecraft, for high density computer data storage, and for high speed computer printers. Lasers are also used in many other engineering areas such as surveying, security systems, sporting systems (pitcher and batter training), graphics, process monitoring and control, materials processing, and remote sensing. Some of these are discussed in the next section as industrial applications.

OPTICS AND PRECISION MEASUREMENT

The Space Telescope. Using lasers to form interference patterns is the most precise method we have for measuring short distances and surface contours. Laser-generated interference patterns have been used to perfect the largest, most precise object ever manufactured, the 2.4-meter-diameter primary mirror of the Hubble Space Telescope. The space telescope represents the largest single advance in astronomical instrumentation since Galileo invented the telescope. It is designed to orbit in space 550 kilometers (300 miles) above the earth, where atmospheric absorption, atmospheric turbulence, city lights, and air pollution cannot affect the quality of light received from the rest of the universe. The space telescope has the capability of allowing us to see seven times deeper into space, to detect objects fifty times fainter, and to see objects ten times more clearly (ten times the resolution) than we ever have before. This means that we can explore a volume of the universe 350 times greater than we have been able to probe in the past. Scientists expect to discover many new things about the universe, to solve some important outstanding questions in astronomy and cosmology, and to uncover some new problems to solve.

In this artist's conception of the Hubble Space Telescope in orbit, the telescope is flanked by two large solar panels that generate electric power. Communications antenna dishes are shown *above and below* the telescope structure.

(Drawing courtesy of Lockheed Missiles & Space Company, Inc., Sunnyvale, California)

There is another important reason why the space telescope can greatly enhance our knowledge of the universe. The atmosphere absorbs a great many different wavelengths, especially those in the ultraviolet and the infrared regions of the spectrum. Ground-based telescopes can receive wavelengths in a range of only about 300 to 1,000 nanometers. The space telescope is limited only by the response capability of its sensors, which is from 115 nanometers to about 1 millimeter. This is a thousand times the earth-based wavelength range.

The space telescope is the size of a railroad boxcar. It weighs 11,600 kilograms (25,500 pounds), is 13.1 meters (43.5 feet) long, and is 4.3 meters (14.0 feet) in diameter. The telescope's optical system was built by Perkin-Elmer Corporation, and the spacecraft housing, power system, positioning and control system, and communications system, by Lockheed Missiles & Space Company, Inc. It is a Cassegrain-type telescope: a secondary mirror reflects the incoming light back through a hole in the center of the primary mirror, where it arrives at the focal plane to be received by five different types of scientific instruments. The data and images developed by highly sophisticated instruments are digitized and transmitted by radio back to earth for analysis at the Space Telescope Science Institute at Johns Hopkins University. The telescope spacecraft housing has one of the most precise positioning systems ever developed. It uses gyroscopes, reaction wheels, and fine-guidance star sensors to hold the position of the telescope onto astronomical objects to within 0.01 arc-second over a 10-hour period. This is about 3 millionths of a degree. (A second of arc is 1/3,600 of a degree.)

Lasers have played an important role in the creation of this valuable instrument. During the polishing of the mirror, which determines its final shape, a helium-neon laser was used with various other optical components to form an interference pattern of the mirror's surface. The interference pattern enabled the polishing engineers to keep track of the shape of the surface to within about 25 nanometers (a millionth of an inch). Information from the interference pattern was fed to computers during each stage of polishing to generate topographic maps of the surface of the mirror. These maps indicate to the

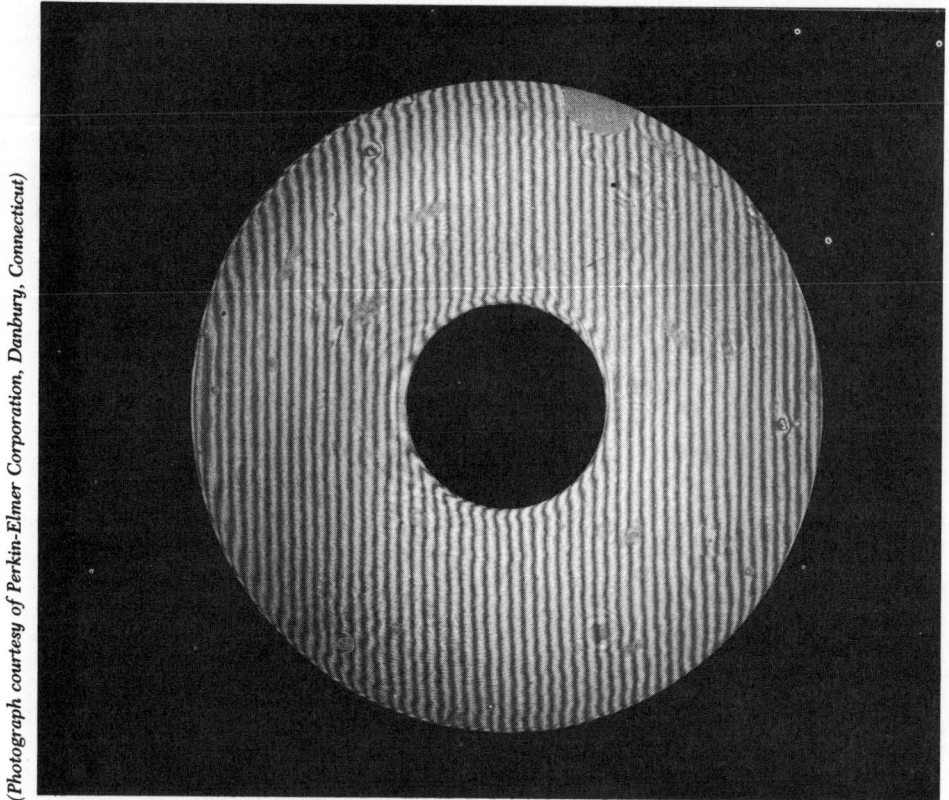

This interferogram of the space telescope primary mirror results from the mixing of laser light reflected from the mirror surface with a reference beam in a specially designed large-aperture interferometer. A helium-neon laser is used to form the beams. The interferograms are scanned by microdensitometers and computer analyzed to derive surface contour and aberration data. Resolution of surface data is to within less than a nanometer.

engineers which parts of the surface are above, at, or below the required hyperbolic shape that the mirror must have in order to precisely focus the light and form clear, undistorted images of the objects under study. The mirror's surface is so perfect that, if the mirror was expanded to 5,000 kilometers (3,000 miles) in diameter, its surface would vary by no more than 6 centimeters (2.5 inches) from a perfect hyperbola.

The Ring Laser Gyroscope. Inertial navigation systems are

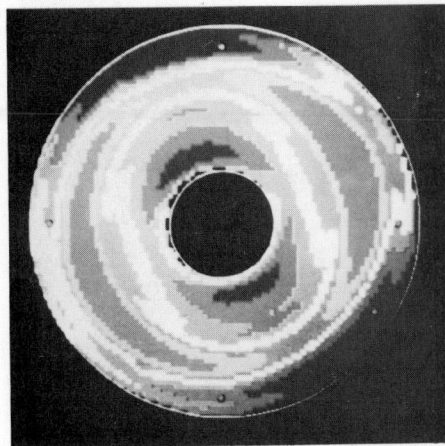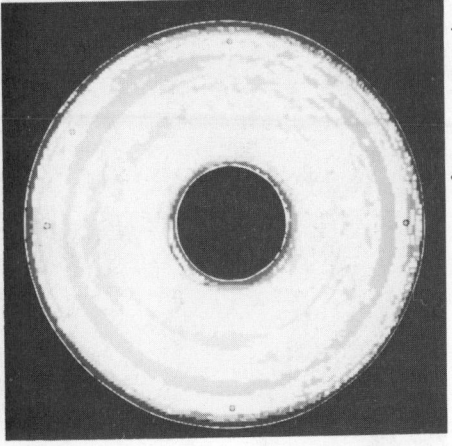

(Photographs courtesy of Perkin-Elmer Corporation, Danbury, Connecticut)

These topographic maps are derived by computer processing of laser interferograms of the space telescope mirror surface. They show how much the surface is changed by the computer-controlled polishing. The first map, taken before polishing, shows the high points in blue and the low points in red. These represent deviations of a few micrometers from the ideal hyperbolic surface. The second map shows the mirror surface after polishing has been completed. Most of the surface is within the desired limits. The surface deviations have been reduced to within a few tens of nanometers.

used in ships, aircraft, missiles, and spacecraft to measure their acceleration and determine their location and velocity. The heart of an inertial navigation system is the gyroscope, which senses acceleration. Signals from the gyroscope are sent to computer microprocessors, which derive the velocity and location. Results of the computations are sent to displays for pilots, relayed to control systems to guide the craft to its destination, or relayed back to a control station. These systems are very important to the military, commercial aircraft companies, and space agencies.

Conventional gyroscopes are mechanical and do their job very well, but they are very sensitive and take a long time to warm up and stabilize. They are subject to temperature, magnetic, and gravitational disturbances. Gyroscopes are now being made using lasers. The laser gyroscope measures rates

(Photograph courtesy of Singer Kearfott Division, Little Falls, New Jersey)

This ring-laser gyroscope contains an optical system consisting of six mirrors forming three square resonators with their planes perpendicular to each other. This enables one gyroscope to sense movements in all three dimensions at the same time. The single block of low-expansion material is filled with a helium-neon mixture and excited by a high-voltage DC discharge. This results in laser action that sends one wave propagating clockwise and another wave propagating counterclockwise in each of the three resonators. The interference of the waves produces a standing wave pattern that is stationary in space, regardless of any motion of the device itself. As the apparatus is rotated, the standing wave pattern remains stationary while the sensing mechanisms pass through the pattern, generating pulses detected at each node of the pattern. These pulses are the output of the instrument which is sensitive to angular motions of approximately 2 arc seconds.

of angular movements. These signals can be used to derive the same type of information as is provided by conventional gyroscopes.

The laser gyroscope consists of a ring laser, in which the resonator forms a loop reflecting the beam along a path with a shape that is an equilateral triangle or a square. When the resonator of a laser is in the form of a ring or loop, two traveling waves are set up, one traveling in each direction around the loop. When the gyroscope is rotated, the frequency of the light moving in one direction is changed with respect to the frequency of light traveling in the other direction. A small part of each beam is transmitted from the resonator by partially reflecting mirrors, and the beams are mixed together to form an interference pattern of fringes, much like those shown in figure 4.7. The position of the interference pattern is sensed by very sensitive detectors whose signals are sent to the microprocessing equipment.

Laser gyroscopes have a large number of advantages over conventional mechanical gyroscopes. They are more stable and can be used immediately, without a long warm-up period. With no moving parts, they are insensitive to temperature, magnetic, and gravitational effects. They are more rugged and reliable. A typical conventional gyroscope has a mean time to failure of 1,000 hours; laser gyroscopes have achieved 32,000 hours. The laser gyroscope is not affected by high accelerations that occur during aircraft or missile takeoff and maneuvering.

In 1982 a new method of detecting movements of the interference pattern from a laser gyroscope, using charge-coupled imaging semiconductor sensors, resulted in an improvement of laser gyroscope resolution by a factor of 100. The minimum detectable angular movement, using nonimaging single-point detectors that were originally used in laser gyroscopes, was 2 seconds of arc. The new detection method provides a resolution of 0.01 second of arc. This improvement makes the laser gyroscope 10 to 50 times better in angular resolution than conventional gyroscopes. The engineering field of guidance and control is just one of many fields that have been radically changed and improved by use of laser technology.

COMPUTER ENGINEERING

Optical Data Storage. Beginning with development of the laser video-disk in the early 1960s, scientists and engineers became interested in the possibility of optically storing and retrieving large volumes of computerized information using small devices. Optical techniques offer the possibility of storing 17 to 100 times as much information in the same space as can be stored with the magnetic methods used in floppy and Winchester disk drives. A 13-centimeter (5¼-inch) optical disk can hold 100 megabytes of information, as opposed to a 13-centimeter floppy disk that holds 350 kilobytes of information. A byte is 8 computer bits of ones or zeros, enough to store one character of digital information.

Besides increased storage density, optical systems have the advantage of reading and writing without contact with the disk. This eliminates surface wear and the head-crash problems of magnetic disks. At the present time, the biggest disadvantage of optical data storage is that a disk can be written only once and thus is not erasable. If the data on the disk is to be changed, a new disk must be written. Corporations and research agencies throughout the world are intensively searching for an erasable optical material.

The increased density of information storage opens up a whole new realm of possibilities for computer applications. An optical card the size of a credit card is available that can be carried by any person and contain his or her entire medical history, including X rays. Doctors can place the card in a computer and instantly have a complete set of medical records. A card like this holds 2 megabytes of data. Companies and government agencies with large-scale information storage requirements, such as insurance companies, which have been storing thousands of computer magnetic tapes, are now placing all their historical information on a few optical disks that can be stored in one cabinet. There are additional applications for optical disks in systems requiring large archival storage of data, such as earth-resources satellite imagery, medical records imagery, and the Defense Mapping Agency's detailed mapping of all land masses on earth.

Commercial optical-data disk drives are now available that can store 1,000 megabytes on a 30-centimeter (12-inch) disk, the equivalent of 10 reels of ½-inch 6,250-bit-per-inch tape or 500,000 pages of printed characters. The system uses a 20-milliwatt gallium aluminum arsenide diode laser focused to less than a micrometer to write 5,700 bits per centimeter

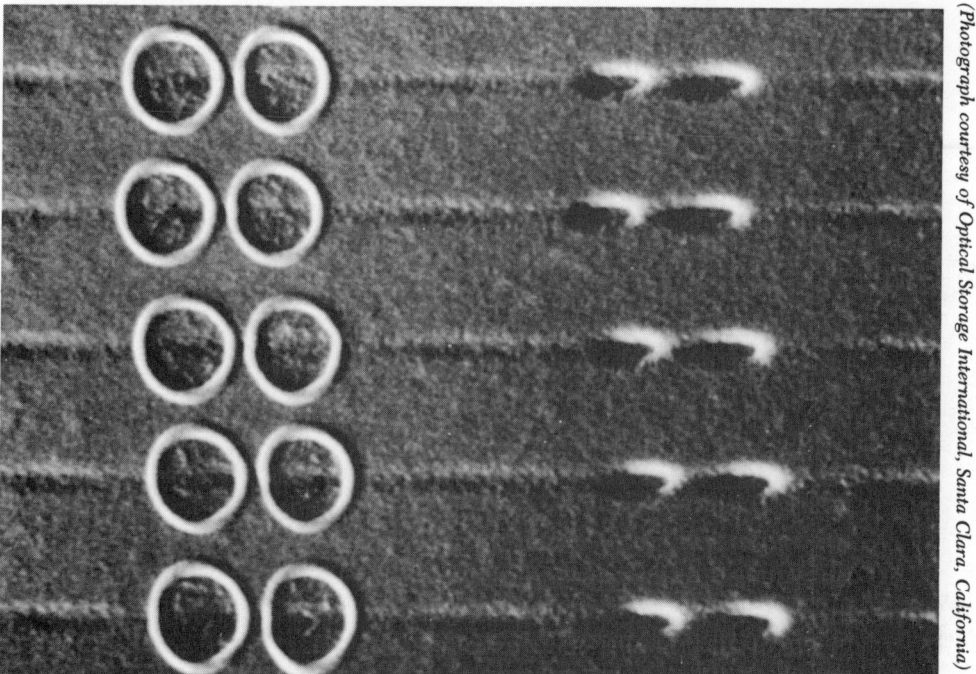

(Photograph courtesy of Optical Storage International, Santa Clara, California)

This is a scanning electron microscope of the surface of an optical disk. Pits formed by the diode laser are on the *left;* they are a little more than one micrometer in diameter. Each pit represents a computer bit and is detected by its lower reflectivity compared to the surrounding surface. The horizontal grooves and the indentations on the *right* are prestamped on the disk and are used for positioning and timing. A 30-centimeter (12-inch) disk stores 1 gigabyte of computer data.

(14,500 bits per inch). Each bit is a small pit about one micrometer wide and slightly more than one micrometer long in an optically sensitive material. The holes are small enough that strips of bits called tracks can be placed within 1.8 micrometers of each other, permitting 5,700 tracks per centimeter. The density of data storage is 33 megabits per square centimeter, which is 700 times that of the floppy disks currently in use.

All this is made possible by the fact that a high-quality diode laser can be focused to a spot less than a micrometer in size and can be modulated at a rate of a few tens of megahertz. The writing of data into the storage media requires 10 to 20 milliwatts of power. Diode lasers of this power and beam

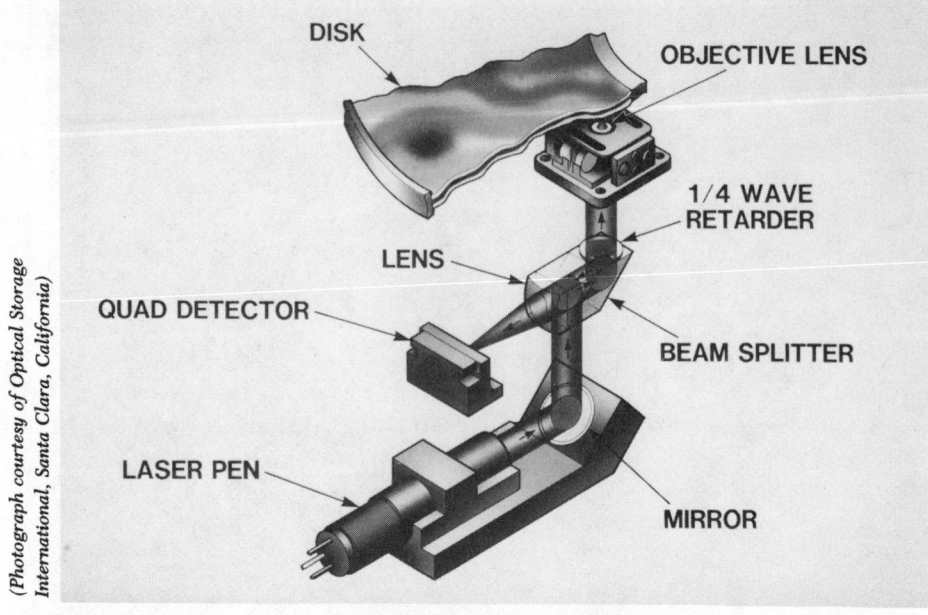

DISK

OBJECTIVE LENS

1/4 WAVE
RETARDER

LENS

QUAD DETECTOR

BEAM SPLITTER

LASER PEN

MIRROR

(Photograph courtesy of Optical Storage
International, Santa Clara, California)

The heart of the digital optical disk drive is the carriage light assembly. The diode laser is in the laser pen shown at the *lower left*. Its light is carried up the light path and is focused on the disk. Light reflected from the disk is intercepted by a beam splitter and directed to a quad detector that sits just *above* the laser pen. The quad detector makes it possible to read the disk at the same time it is being written, in order to verify that the write is correct. Each disk is written once and can be read any number of times.

quality are not easy to come by. Since they must be selected, by virtue of their beam quality, from large production quantities, they are relatively expensive. They usually operate in the 780- to 830-nanometer wavelength range. The most common write-once and read-only storage media are films of tellurium alloys. Information is read back with only fractions of a milliwatt. Research on erasable optical storage is based on two types of materials. The first type, magneto-optical, is read out with the assistance of a magnetic field. Data storage in the material changes the polarization of the reflected light. The second type, materials that change phases that differ in reflectivity, change from amorphous to crystalline and back to amorphous states in response to laser light. The data is detected as changes in reflectivity of the material. With the perfection of erasable materials, optical data storage will go into head-to-head competition with magnetic data storage, replacing the older technology in many applications.

This same technology is making possible the rapid entry of audio compact disks into the market place. Audio compact disks are only 13 centimeters in diameter and contain more high fidelity music than a full-size, long-playing record. As prices of compact disk players and compact disks come down, they will become more popular. The disks are written and copied at the recording studio and played back in the home by compact disk players using low-power diode lasers. The fidelity of compact disks is far superior to that of records. Noise from imperfections in record surfaces is eliminated. The disks do not wear out as records do, because there is no direct material contact on the disk. Because of the method used to encode the audio information on the disk, smudges and dirt on the disk do not interfere with performance. Here is yet another example of a new laser-based technology that is most likely to eventually replace an old technology.

Laser Printers. The combination of a helium-neon or a diode laser with an acousto-optic or an electro-optic modulator and a scanning device such as a rotating polygon shaped mirror forms a *laser scanner*. Laser scanners are used in many applications to generate or read out images and character data. They are used in high-speed nonimpact printers, marking systems, range finders, laser radars, phototypesetters, barcode readers (the laser checkout reader in your supermarket), optical inspection systems, photolithographs, optical character recognition systems, ophthalmic imagers, and robotic sensors, to name just a few. Laser scanners provide high-quality, high-speed, reliable capabilities that are very cost competitive with other types of systems. An excellent example of a use of laser scanners is the nonimpact laser printer.

A simplified block diagram of a laser printer is shown in figure 7.3. A laser printer begins with an electronic system that takes character data from the computer and transforms it into a video signal that defines the image of the characters on a dot-by-dot basis. This signal is fed to the modulator, which deflects the beam depending on whether or not a dot is present. The laser beam is scanned across a rotating drum using a scanning mirror. The movement of the mirror forms the image across the page, and the movement of the drum forms the image down the page. The drum consists of a photoconductive material and is electrostatically charged. Regions where the laser beam hits to form a dot are made conductive and lose their charge; regions where the beam does not hit retain their charge. A toner (ink), which consists of a fine powder of electrically charged particles, is brought near the drum. The toner adheres by electrostatic attraction to the areas of the

Figure 7.3. Block Diagram of a Laser Printer

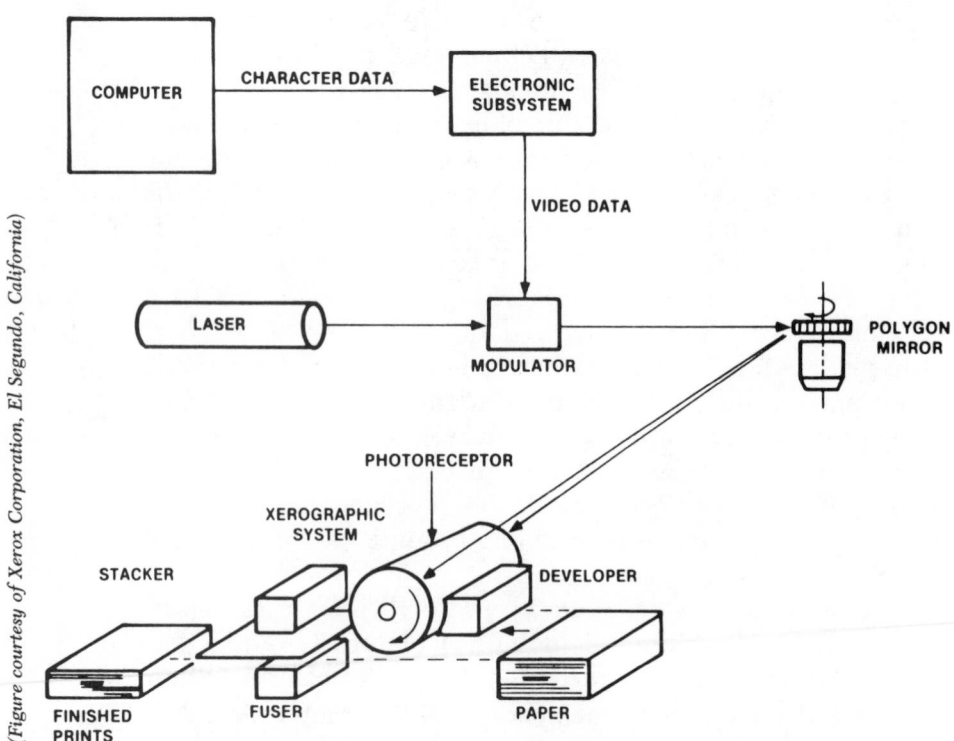

(Figure courtesy of Xerox Corporation, El Segundo, California)

drum that are charged. As the paper is brought near the drum, a charge is applied to the paper, causing the toner to move from the drum to the paper. The toner is fixed on the paper by heat, which melts it into the paper. Toner left on the drum is cleaned off by a scraper to prepare it for the next page.

Today's high-speed laser printers are capable of printing 100 to 150 pages of 8½" by 11" paper per minute. The quality of the print is vastly superior to the old impact printers, and many different type faces and special characters are available. As developments continue, the cost of printers is coming down and reasonably priced models for use with personal computers are becoming available. Newer models use diode lasers to replace helium-neon lasers, which considerably lowers cost by saving space, simplifying modulation, reducing warm-up time, and reducing the frequency of repairs. The use of laser printers is rapidly expanding as they replace obsolete models and are sold with new computer systems.

Industrial Applications of Lasers

Lasers are being used extensively in industry for manufacturing, materials processing, inspection, process control, chemical synthesis, and special applications. Materials processing consists of drilling, cutting, welding, surface treatment, and machining assistance. Most applications are in the areas of metals and semiconductors, but plastics, ceramics, glass, rubber, composites, paper, and wood are also being processed with lasers. The number of uses of lasers in industry is increasing daily. High-power lasers and the processing and control systems that go with them are expensive, complex, and require special expertise to use. Why, then, use lasers? The primary answer to that question is that many jobs can be done much better using lasers. More and more ways are being discovered for lasers to do a job faster, more efficiently, and at less cost than conventional methods. Indeed, lasers are now making possible the manufacture of many parts and components that previously could not be made.

MATERIALS PROCESSING

Parts are now being designed to take advantage of the advanced processing capabilities of laser systems. One example is the combination pulley and clutch plate for automobile air-conditioning compressors that is produced by a major American automobile manufacturer. This part weighs half as much as previous parts, saving weight without sacrificing performance. Its parts contain joints that could not have been welded by ordinary methods. The welds are made by three 5-kilowatt continuous CO_2 lasers. The weld melt, one-quarter inch deep, fully penetrates the contact required between the parts. A complete weld is done in 10 seconds. Production of this part is expected to eventually reach 30,000 per day.

Other methodologies compete with laser technology, but lasers are frequently the most desirable method. Electron-beam welding offers capabilities similar to laser welding, but electron-beam systems are difficult to maintain, and the parts being welded must be in a vacuum. Electron-beam systems are typically out of service 60 percent of the time compared to 20 to 30 percent down time for laser welders.

CO_2 and Nd:YAG lasers almost totally dominate laser materials processing such as cutting, welding, marking, and annealing. The CO_2 laser is the most common industrial laser, especially for very-high-power applications. For purposes of cutting and welding heavy steel, CO_2 lasers in the 20- to 30-kilowatt region are needed. These power levels can handle steel up to 2.5 centimeters (one inch) thick. The most powerful currently in use is the 25-kilowatt CO_2 laser built by Avco-Everett and being used at the Westinghouse Laser Center in Sunnyvale, California. A laser of this capability with its associated equipment and power supplies is large enough to occupy a small building. Majestic Laser Systems, in association with the Electrical Engineering Department of the University of Alberta, is applying advanced excitation techniques to produce more compact, lower cost, easier to control commercial CO_2 lasers with 20 kilowatts capability. For applications requiring less than one kilowatt, either CO_2 or Nd:YAG lasers may be used. The choice depends on the particular application's requirements, the capabilities of the lasers, and their associated control systems and cost.

Lasers do not stand alone. There must be a way of delivering the laser light to the workpiece, controlling the delivery, and controlling the process. Two very important devices make these functions possible: the microcomputer and the robot. Specially programmed microcomputers are an important part of most industrial laser systems. They are used to control and time the laser and the parts being processed. The parts must be moved into place, the laser must be activated, the part or the laser beam must be moved in the proper pattern and with the necessary precision to accomplish the processing, and the laser must be deactivated and the part moved out so that the next one can come into place. Robotic arms and various positioning devices are controlled by the microcomputers and used to move the laser beam or move the part to get the job done. The robot arm consists of an elaborate assemblage of electronics, sensors, motors, and mechanical devices. The arms are capable of intricate and precise maneuvers. As the robotic arms move to engage the workpiece and keep the laser energy focused at just the right position, the laser energy must be directed along the robotic arm. For applications requiring less than 20 to 30 watts of average power, fiber-optic cables can be used. Fiber optics cannot be used for CO_2 lasers because the glass they are made of does not transmit 10.6-micrometer radiation. For CO_2 lasers with less than 30 to 40 watts of output power, a flexible, internally reflective tube called a *waveguide* can be used. For higher-power applications, reflective prisms inside articulated optical arms must be used to deliver the laser energy.

When very large or heavy objects are being processed, it is easier to move the laser beam. When the object is small and needs to be processed on two sides or around a circumference, it is easier to move the object. At other times, it is advantageous to move both the laser beam and the object being processed, as in the case of continuous production where the material is running through the laser system and the beam is moved back and forth to cut or weld in a prescribed pattern. Clothing manufacturers are using systems in this manner to cut patterns in many layers of cloth at the same time. The laser makes a clean, neat cut that is precisely controlled and does not disturb or move the cloth.

A recently developed 800-watt CO_2 laser is one-tenth the size of other lasers with output in the range of 500 to 1,000 watts. It may be suitable for mounting directly on the robotic arm, reducing the complications and losses of the optical delivery system.

Laser processing of materials is usually employed in situations where there are special requirements or where processing by conventional means is difficult or produces poor results. An excellent example of an area where lasers have found extensive use is in the manufacture of extremely small devices (microfabrication). The laser light can be focused down to a few tens of micrometers, and its position can be controlled to within less than a few micrometers. This means that very small objects, some requiring a microscope to be seen, can be drilled, cut, and welded with incredible precision. Holes and cut surfaces are smooth and clean, and do not require deburring. The laser is especially valuable for processing very hard substances such as hardened steels, nickel alloys, and titanium alloys.

Welding. Lasers can weld everything from microcircuits to heavy steel. The largest welding lasers are capable of outputs of 20 to 30 kilowatts continuously. Laser welding requires no filler material and minimizes distortion, since it heats only at the exact point of the weld and makes no direct contact with the part, thus applying no pressure. Localized heating often results in stronger, more reliable welds, and the bead placement is very accurate. For example, the blades of trephines, devices used in eye surgery, are welded onto their handles by laser, and the cases for cardiac pacemakers are welded together by lasers. Each of these instruments requires precisely controlled, reliable welds. The United States Navy has built a huge welding facility to aid in shipbuilding. This facility is capable of welding huge ship components made of heavy steel. The facility is saving money in the cost of building ships because it can weld faster and with better quality than can be done by hand.

The French have developed a television camera and image-

(Photograph courtesy of Coherent-General, Palo Alto, California)

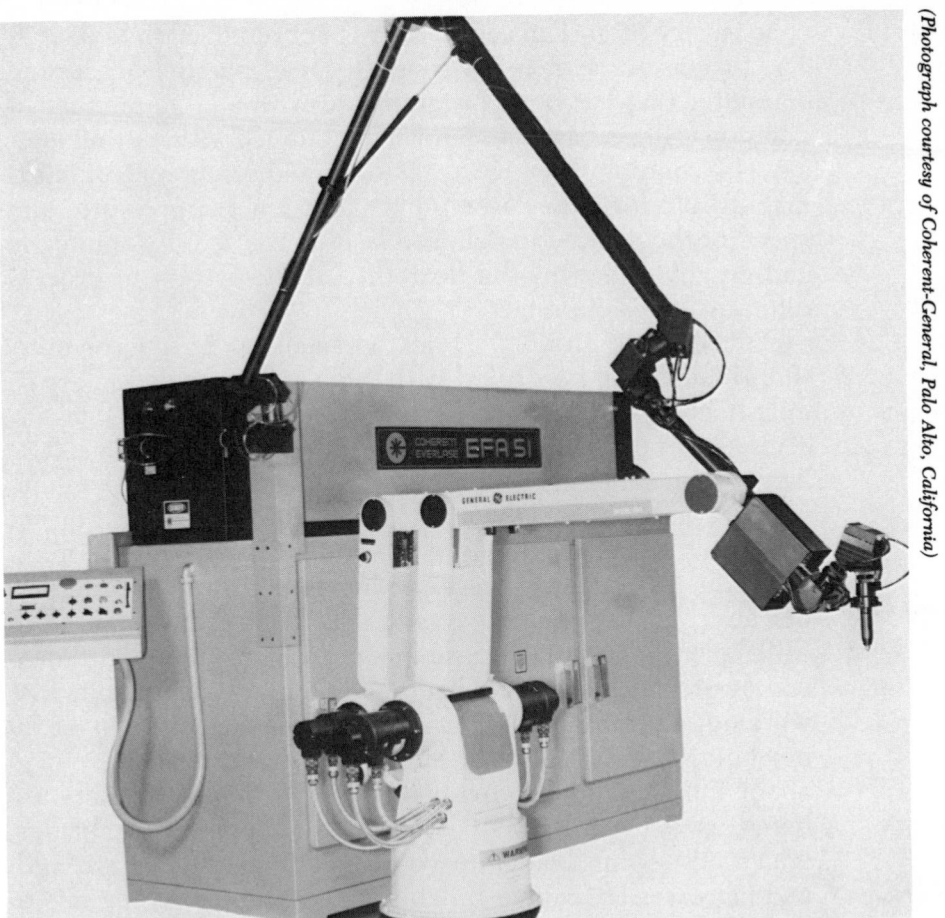

This photograph illustrates the use of a robotic arm to control a high-power cutting and welding laser. The high-power CO_2 laser, manufactured by Coherent Everlase, delivers its 3,000-watt beam through the articulated black light pipe to focusing optics in the tip of the instrument that looks like a large ballpoint pen. The General Electric P50 robot is microcomputer controlled and moves the laser output in a preprogrammed sequence of motions to accomplish the required welding or cutting job. The arm also keeps the laser beam precisely spaced from the working material to maintain uniform cutting or welding action.

processing system to automatically control cutting of metals by an oxygen-assisted laser cutter. The camera picks up images of the shape and position of the cone of incandescent metal particles that emerges from below the melted metal. Since the cone is very bright, an active laser is not required to look at it. The images are sent to a very-high-speed parallel processing computer that has been programmed with a large number of decision criteria based on human experience with cutting metals. The computer makes its decisions and sends control information to adjust the laser power, oxygen gas pressure, and speed of the moving metal. The laser cutter is fully automatic and greatly enhances the flexibility of the system to process different types of metals.

Cutting and Drilling. Holes as small as 25 micrometers (0.001 inch) can be drilled with Nd:YAG lasers and holes 75 micrometers (0.003 inch) can be drilled with CO_2 lasers. These holes can be placed in materials too hard or too soft to be cut with drill bits. Examples of such materials are titanium, plastics, and rubber. One of the first industrial applications of lasers was to drill the holes in the ends of baby bottle nipples. In production systems, holes can be drilled at rates of millions per hour. In the automotive industry, lasers are being used to drill holes in internal transmission gears made of hardened carbon steel. This process had been done by carbide bits that wore out after only six or seven holes; the cost came to $1.50 per hole. The cost of laser cutting the holes averages out to about $0.10 per hole. In the aerospace industry, lasers are being used to drill small holes in titanium turbine blades, which allows the blades to cool without sacrificing strength and increases jet engine reliability and lifetime. Using bits to drill such small holes in material this hard would be nearly impossible and very expensive. Lasers are even being used to cut diamonds, one of the hardest natural substances.

Lasers are especially useful for cutting complex shapes in materials that are either difficult to machine or might easily be damaged. Using lasers eliminates the need for dies and eliminates stress and distortion induced by mechanical stamping. Other advantages to laser cutting are the elimination of tool wear or slippage and minimum heating of surrounding areas. In addition, the cutting quality and rates are not dependent on material hardness. An excellent example of the laser's usefulness is in production of titanium wire-mesh pads used for joining artificial knees to leg bones. A mesh of such hard wire would be enormously difficult to produce by mechanical means.

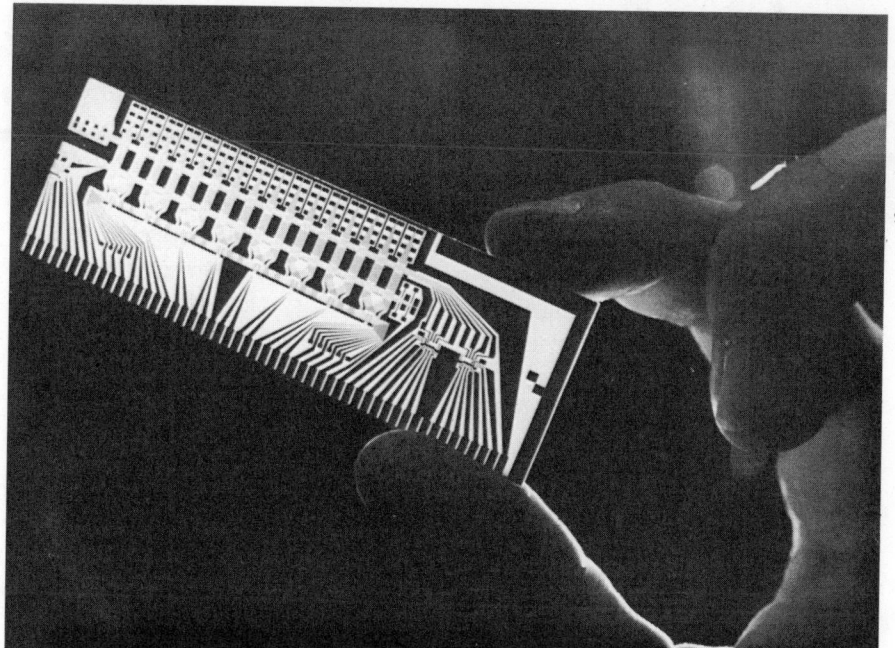

(Photograph courtesy of AT&T Bell Telephone Laboratories, Murray Hill, New Jersey.)

This miniature circuit pattern was machined by laser light directly onto an insulative backing material. Circuits such as this one are the building blocks of sophisticated electronic components used in switching and transmitting devices for telephone equipment.

Another excellent example of the usefulness of lasers for cutting lies in the production of sawblades. Conventional methods of sawblade production require many steps, each involving handling of the blades. Custom blades and parts cannot be produced without expensive retooling. A computerized design and control system working in conjunction with two 700-watt CO_2 lasers is handling 70 percent of a major sawblade company's production with a 65-percent reduction in production time. The blades, produced in a single step, are cut from sheets of steel 0.378 centimeter (0.149 inch) thick. The laser beam is focused down to 76 micrometers. The accuracy of the contours of the blades is approximately 130 micrometers (0.005 inch). Production of custom blades depends only on entering the necessary parameters into the control computers. The company has significantly increased production of custom blades.

Surface Treatment. High-power-density lasers focused and scanned across heat-treatable materials such as carbon steels can significantly increase the hardness of manufactured parts. The laser momentarily melts the surface, which quickly cools and resolidifies. Laser hardening of medical instruments has increased the service life of most instruments by 2.5 to 3 times. Surface hardening is also very important in the manufacture of critical engine components. Hardening the most wear-prone components can significantly increase the time between rebuilds, thus reducing engine operation costs. A major manufacturer of large diesel engines for railroad, marine, oil rig, and utility applications is laser hardening the engine cylinder bores. The laser method was chosen over several other surfacing and treatment methods; although expensive to implement, it has proven effective and reliable.

Besides surface hardening, lasers are also being used for

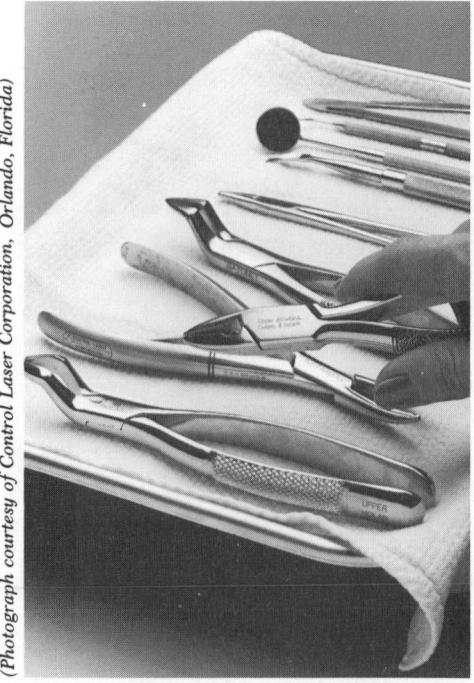

(Photograph courtesy of Control Laser Corporation, Orlando, Florida)

The markings on these dental instruments were made by a 50-watt continuous-wave Nd:YAG laser. Two scanning mirrors controlled by a microcomputer move the laser beam to position and form the characters. A lens focuses the beam on the workpiece. Characters are written at speeds of 20 centimeters per second (8 inches per second). They can be of different sizes and different shapes, and can have different line thicknesses. The characters can be written on straight lines or in circles, and can be placed on flat or curved surfaces. Extraordinarily small objects can be clearly marked. The marking system, called InstaMark, is used to engrave metals (steel, aluminum, titanium, tungsten), ceramics, silicon, Lexan, polycarbonates, and plastics. Since no pressure is applied to the workpiece, delicate parts can be marked without damage.

metal cladding (covering the surface with another type of metal), alloying (mixing metal with other metals), and particle injection. Particle injection involves momentary melting of the surface so that high-melting-point particles such as carbides can be embedded in the surface. Melting and rapid recooling of many materials can result in formation of especially smooth surfaces with superior physical properties. Surface treatment of this type is very important for the aerospace industry.

Marking. Because lasers can precisely remove very small quantities of material from a surface, they are valuable for marking and identifying manufactured objects. Laser marking is especially useful for materials that are very hard and thus very difficult to mark, such as stainless steels like those used to make surgical instruments, titanium used to make aircraft parts, and precious gems. Laser marking is computer controlled and can be done at very high speeds. As little as 25 micrometers of material is removed from the surface to make the mark. Besides labeling medical instruments and objects like titanium pacemaker cases, lasers are being used to identify precious gems and jewelry. No matter how hard the surface, the laser can inscribe a complete set of identifying information in an area so small it takes a microscope to read it. The value of gems is not reduced by such marking, since gemologists do not consider it a defect.

Soldering. Computer and aerospace companies are turning to the use of lasers to solder circuit boards. With increased microminiaturization, circuit boards are becoming smaller and more complex and thus more difficult to handle and solder by hand or even by soldering machines. Computer-controlled Nd:YAG or CO_2 lasers can be precisely directed to the right locations and can be made to provide just the right heat level to solder circuit boards. Boards are soldered in one-tenth the time it would take to do it by hand. Reworking due to defects has been reduced by 90 percent. Soldered joints are very uniform and reliable. Considerable savings are being realized by reducing labor costs and at the same time making the work performed more reliable. Purchasing, setting up, and starting the operation of a laser soldering system is initially very costly, but it can pay for itself through cost savings within a few years.

Semiconductor Processing. Semiconductors are one of the most important production areas of high technology. They are the basis material for all diodes, transistors, and integrated circuits in nearly all modern electronic equipment. Semiconductors must be formed, cut, etched, annealed, and impregnated with impurities, all on a microscopic scale and with a high degree of cleanliness. Nd:YAG and especially excimer lasers are being used to refine and improve the processing of semiconductors.

Because of their higher power and lack of spatial coherence, excimer lasers are ideal sources of ultraviolet radiation for the photolithographic processes that form semiconductors into integrated circuits and into microcircuits. The Nd:YAG laser can be used to anneal semiconductors. During the annealing of a semiconductor, the surface is momentarily melted. As it cools, the silicon regrows into its cyrstalline structure. The regrowth improves purity and removes damaged areas, improving the overall quality of the semiconductor material.

A prime advantage of laser processing is that it can be carried out with the semiconductors isolated from the outside world to maintain complete cleanliness. The laser energy is controlled, moved about, and directed through windows into the semiconductor working area. Semiconductors can be selectively converted into P-type and N-type materials by making the correct impurities available in the form of gaseous compounds surrounding the semiconductor surface. As the laser carefully melts the semiconductor to a prescribed depth, the impurities are incorporated into the crystalline structure.

In addition to conventional photolithographic processing, lasers are being used to directly pattern semiconductors by photodecomposition of the surface and initiation of photochemical reactions that can deposit materials or etch away the surface. Photodeposition of conducting material such as gold onto integrated circuits is used to repair or customize them. The materials are deposited when the laser breaks down gaseous compounds whose molecular structure contains the metal being deposited.

Lasers can also be used to directly form and trim the semiconductor patterns that make up individual circuit components such as resistors and capacitors. The lasers are also being used to form the intricate photomasks required to indicate where semiconductor processing is to occur. The masks control the location of material deposition, diffusion of impurities, and etching away of surfaces.

The laser is making it possible to even further reduce the size of electronic components. Miniaturization makes it possible for the Space Shuttle to carry its own computers, for aircraft to find their way when visibility is low, and for computers to be incorporated in watches. Soon it may be possible to carry complete communication systems on our wrists, just as Dick Tracy and Captain Kirk have been doing for years.

INDUSTRIAL PROCESS MONITORING, INSPECTION, AND CONTROL

Industrial process monitoring, inspection, and control is a rapidly expanding area of laser application and is of great importance to industry and manufacturing. Laser-based systems are being designed and used to perform many tasks that previously could be done only by human beings, jobs that are repetitive and difficult for workers to do for long periods of time. In many cases, the laser system can do the job faster and more reliably than a human being, contributing to significant improvements in the quality and speed of industrial production.

Two primary types of systems are used for industrial monitoring, inspection, or control. These are video-based (television) systems and laser-based systems. Each type of system has advantages and disadvantages for particular production situations. The video-based system consists of a television camera to capture images of parts under manufacture. The camera is coupled by analog to digital converters that feed the images into microcomputers. The computers analyze the images and make decisions on action to be taken. The actions may consist of sending control signals to manufacturing equipment or putting out signals to inform workers of the quality, type, or dimensions of materials or parts in production.

The laser-based systems consist of the laser, a scanning device to move the laser beam across the objects being investigated, and an imaging optical system to analyze the pattern of reflections from the object. Since only low-power, inexpensive lasers are required, these systems usually consist of diode or helium-neon lasers. The scanners can be electro-optic or acousto-optic. They are usually mechanical, consisting of various types of rotating mirrors or rotating disks with polished facets. The imaging system analyzes the characteristics of the beam spot or pattern of reflections from the object. In simple systems, decisions can be made by direct-analog electronic means, but in more complex systems, the data is digitized and sent to a computer.

Video systems tend to be slower, requiring more time for image processing. For simpler applications, the laser systems are very fast and can more easily detect edges and object features. The laser systems can discriminate between reflected and diffusely scattered light, whereas the video systems see only scattered light. For smooth materials, such as plastics or metals, laser scanning is superior. The quality of a surface can easily be determined. Roughness and imperfections such as holes can be detected much more easily by laser systems. In general, more complicated tasks are better done by video systems, while simpler tasks and tasks requiring higher speed can be more effectively done by laser systems. In theory, however, laser systems can be designed to do more complex tasks and will always have the advantage of having their own source of light, allowing a much greater facility to detect the reflected or scattered light for analysis.

Many interesting control systems, both laser and video based, have been developed to control laser and conventional processing applications. We will discuss three of these as interesting examples of automation developed because of lasers.

A pulsed Nd:YAG laser is being used to measure the quality of solder joints. It can even look for internal voids, places where there is not enough solder, that human inspectors cannot see. The quality of solder joints in printed circuit boards is of great concern, not only to all electronic equipment manufacturers but especially to the military and space agencies, which require highly reliable electronics. The laser is focused onto the solder joint and fires a 30-millisecond pulse. An solid-state detector picks up infrared radiation given off by the heated joint. The manner in which the joint cools will reveal whether or not the joint is good, has too much solder, has too little solder, or has an internal void. If the joint is bad, the circuit board is sent back for reworking.

Laser systems are especially useful for making measurements in hostile environments like steel manufacturing, where temperatures would destroy material probes. A gauging probe consisting of a pulsed diode laser, a charge-coupled-device imaging camera, and a microcomputer can measure the thicknesses of hot slabs in reversing mills or the level of molten steel in casting operations. The system can also be used to inspect hot pipes and tubes and to measure the dimensions of forged railroad wheels. This particular laser probe system is able to measure dimensions over a range of 0.8 to 512 millimeters with an accuracy of 0.002 millimeter. The laser has an output power of 4 milliwatts and is modulated at 16 kilohertz. The detector picks up light scattered from the object being measured and can determine exactly where the spot of light on the detector's surface falls. The slightest variation in the position of the surface being measured changes the position of the scattered light on the detector. The output signals from the detector are sent to the microcomputer for analysis and computations. This system is also being used in the manufacture of rubber, plastic, wood, and machine parts.

A laser scanner is being used to inspect for flaws in semiconductor wafers. For example, one type of scanner can inspect 600 wafers per hour and can spot flaws as small as 1 micrometer in diameter. A rotating mirror scans a 5-milliwatt heliumneon laser beam across the wafers. Two sets of fiber-optic cables are used to collect both specularly reflected light (direct reflections as from a mirror) and diffusely reflected (scattered) light. The fiber-optic cables send the light to very sensitive photomultiplier detectors, which generate electronic signals that are digitized and sent to a microcomputer. The microcomputer can decide whether to accept a wafer, specify that it be recleaned, specify that it be repolished, or reject it altogether. The system is able to identify 12 types of flaws. This system is another that replaces cumbersome manual inspections.

We have been able to talk about only a few of the important ways that lasers have contributed to the quality and cost-effectiveness of industrial processes. Many hundreds of other specialized systems have been developed to streamline, improve, speed up, and lessen the cost of methods used in industry.

Applications in Medicine

One of the most surprising, rewarding, and unexpected developments in laser applications has been their use in medicine. A precisely guided and carefully controlled laser beam can remove tumors in areas of the brain too delicate and sensitive to permit conventional surgery. Argon laser beams can be focused on the retina of the eye to weld torn retinal tissue back in place and to coagulate tiny vessels to stop bleeding. CO_2 lasers can be used instead of scalpels to cut tissue, preventing bleeding and guarding against infection. Argon, Nd:YAG, and excimer lasers are being used to remove plaque deposits from coronary arteries. Already hundreds of thousands of people have had their health improved, endured less pain and discomfort in medical procedures, and had their lives saved through the use of lasers in medicine. In the future a large percentage of medical procedures will involve lasers. More ways are being found to apply lasers in situations where conventional treatment is difficult, is impossible, or has severe side effects. A good example is found in ophthalmology, the medical treatment of the eye.

One of the most common medical uses of lasers is in treatment of the retina, the assembly of delicate tissue at the back of the eye that receives light and sends images in the form of neural impulses to the brain. The retina is very delicate and easily damaged. Its texture is like fine tissue paper. Physical manipulation of the retina by instruments must be done very carefully, since there is great risk of damage. However, a low-power argon laser focused onto the retina can spot weld retinal tissue back in place and repair torn, separated, or damaged tissues. The laser can also coagulate small blood vessels to stop bleeding in the retinal area, a process very difficult to effect in any other way. The laser procedure involves no surgery, can often be done in a doctor's office, is quick, effective, and safe, and the patient can usually leave the same day. The patient's comfort is greatly improved, and expensive hospitalization is not required.

For the patient, the use of lasers has simplified medical procedures, increased their effectiveness, and reduced discomfort. For the doctor, laser procedures are simplified and less time is usually required, but he must train to properly use the new equipment. And, of course, the doctor or a hospital must be able to purchase the laser equipment, which, while expensive like other hospital equipment, can reduce hospital operational costs in the long run.

The most common lasers used in medicine are continuously operating CO_2, Nd:YAG, and argon lasers. These are now being supplemented and perhaps replaced by pulsed Nd:YAG, diode, and perhaps most important, excimer lasers. The CO_2 laser is primarily used for cutting tissue, replacing in many cases the use of the scalpel, and for vaporization of unwanted masses such as tumors. The major advantages of laser cutting over conventional means is the prevention of infection and coagulation of the blood vessels. This results in less bleeding and easier visibility for the surgeon, and fewer complications and shorter recovery times for the patient. The advantage of CO_2 is that most tissues have high optical absorption at 10.6 micrometers and are easily vaporized. There are two important disadvantages. CO_2 is difficult to deliver to the operating site. Most fiber-optic materials will not transmit CO_2, and only recently has a special flexible waveguide been developed. Even this has 40 percent losses with moderate bending of a 30-inch section. A second disadvantage of CO_2 is that it cuts by vaporization of the tissue, a thermal process that results in charring and damage to the surrounding tissue. In addition, the CO_2 laser radiation is not absorbed well by the blood and therefore does not allow coagulation as well as other lasers.

The argon laser is also used for cutting and removing of tissue but is even more useful for coagulation to stop bleeding. Coagulation by light is called *photocoagulation*. The red pigment in the blood, hemoglobin, readily absorbes the argon laser wavelength. Most tissues are less absorbent to argon than to CO_2 so that the argon laser light penetrates deeper and is scattered more. The argon laser light can be focused to a smaller spot than can CO_2; thus it is useful in microsurgery, in such areas as the eye or brain. The disadvantages of argon laser light are its low absorptivity for most tissues and the laser's low efficiency. Low efficiency leads to bulky, nonportable systems that require a great deal of input power and cooling. Argon lasers used in some medical applications have up to 28 watts of continuous power; they are devices with large input power and cooling requirements. As better laser devices are being developed, the argon laser is being replaced by the frequency-doubled Nd:YAG, which can produce more power in a smaller, more compact system, can be operated in pulsed mode, and has an output wavelength very close to that of argon. A special advantage of argon and Nd:YAG over CO_2 is that their wavelength of light can easily be delivered by flexible fiber-optic cables.

Nd:YAG can be used in most cutting and tissue removal applications but is at a slight disadvantage because of less absorption in tissues than CO_2. However, it can cause better coagulation than CO_2 and is therefore better for removing highly vascular tumors, tumors with many small blood vessels. New applications are being developed for Nd:YAG when the laser is mode locked and Q-switched. The high-power, short pulses remove tissue more cleanly, without as much burning or damaging of the surrounding area. A mode-locked, Q-switched Nd:YAG laser developed in West Germany is being used for successful treatment of many ophthalmological problems. The laser delivers a train of 4-millijoule pulses of 30-picoseconds duration for a period of 30 nanoseconds. It cleanly and very controllably removes tissue from various areas of the cornea, lens, and iris of the eye.

The latest and perhaps most exciting development in medical lasers is the excimer laser. Of its several advantages over other lasers, one is particularly important. The intense, short wavelength output of excimers can remove tissue cleanly and precisely with no damage to surrounding areas more than a few micrometers from the beam. As we saw in chapter 4, short-wavelength radiation photons have much more energy. When these photons strike the long complex molecules of organic tissue, they have sufficient energy to break up the chemical bond of the molecules, resulting in the production of ionized gases. This process is called *photodecomposition* or *photoablation* and differs primarily from vaporization in that the molecules are decomposed without heating, which would damage adjacent areas.

A few excimer lasers are currently available commercially. Many areas of possible medical use are being investigated in the laboratory. There are two major disadvantages of excimer lasers that must be overcome. The gases they use are hazardous in the operating room, making especially safe containment a must. Another problem is that the ultraviolet radiation may break up the chromosomes in healthy cells, resulting in complications such as tumor growths.

Developments in diode lasers are advancing their potential for use in medicine. Since they are so small, compact, and efficient, medical instrumentation based on them is much easier to handle and operate. They are more stable and do not require constant adjustment by technicians. Diode laser instruments are much smaller and easier to handle. Since they are more efficient, they require much smaller power supplies. It is also much easier and more efficient to deliver the laser energy with fiber-optic cables. Since diode lasers can be as small as a grain of salt, there is even the possibility of sending the laser to the point in the patient's body where it is needed.

With 50 milliwatts available in a single-junction diode and up to 2.5 watts available in junction arrays, microsurgery and microphotocoagulation using fiber optics and endoscopes are coming into use. With more powerful diode lasers being developed, they may soon begin to replace other lasers in medical uses. With other lasers, ideal effectiveness would result from using two wavelengths simultaneously, one to cut and one to coagulate. The diode laser offers the possibility of building lasers that are tailored in output wavelength to the particular type of tissue being processed and, as a result, will be efficient at both cutting and coagulation. Developments in visible diode lasers may make it possible to use them to replace the argon laser, and thus to gain the benefits of using simpler, smaller devices. Pulsed visible diode lasers that operate at room temperature have recently been developed in indium gallium aluminum phosphide with double heterojunctions; they operate at 660 nanometers. However, much development remains before visible diode lasers will be generally available.

One possible application of light-emitting diodes and laser diodes is in the photoradiation therapy of tumors. As greater output powers become available, the convenience of direct delivery to the tumor by fiber-optic cables and endoscopes makes possible much larger numbers of situations where photoradiation therapy can be used against tumors, especially against inoperable tumors. This application requires 25- to 100-milliwatt-per-square-centimeter power densities and 50 milliwatts per square centimeter has been achieved with close-packed light-emitting diodes.

OPHTHALMOLOGY

Lasers are frequently used in the treatment of eye conditions, such as photocoagulation of the retina and treatment of retinal detachment, mentioned above. Many other areas of eye treatment are made possible and enhanced by laser techniques. Besides argon, all the other types of lasers we have mentioned in connection with medicine can be used in some way for ophthalmology. Of the four most interesting areas of application, three are well established and one is just emerging.

The most common use of lasers in ophthalmology is to treat complications due to diabetes. Long-term diabetes without proper control is a major cause of blindness. It results in a condition of the retina called *proliferative diabetic retinopathy*, which includes microaneurysms, hemorrhages, edemas (swellings due to accumulation of fluids), exudates (deposits exuded from the retina), occluded arteries, dilated vessels, new vessel formation (neovascularization), and formation of fibrous tissues.

One of the best treatments for retinopathy is photocoagulation, in which lesions are burned by the heat of a light beam. The argon laser is the most effective laser for this purpose. It can be directly focused to destroy leaky lesions such as microaneurysms and dilated capillaries. One process, called *focal photocoagulation,* is effective in decreasing the rate of vision loss due to a condition called *macular edema.* The macula is the central region of the retina, responsible for the highest-resolution part of vision. In another process called *local photocoagulation,* the laser is applied to local areas or patches of retina, producing burns that destroy neovascularization outside of the optic disk (central retina). This process helps limit the formation of new vessels that interfere with vision. For treatment of neovascularization inside the optic disk, the laser cannot be used directly. However, in a process called *panretinal photocoagulation,* it can be used to place burns scattered over a large area of retina outside of the optic disk. These burns control disk neovascularization indirectly by reducing both oxygen demand and the production of vasogenic factor, a substance that stimulates neovascular growth. Laser photocoagulation is 61-percent effective in reducing severe visual loss. Side effects can occur, but there is less risk than with surgical treatments, and the effectiveness warrants its use.

Another important use of lasers in ophthalmology is in the treatment of glaucoma, a condition in which the aqueous humor (the fluid that fills the eye) fails to properly drain through normal channels. Drainage normally occurs through a ring structure around the iris called the trabecular meshwork, into the anterior chamber of the eye, between the cornea and lens. Glaucoma is first treated medically using drugs to control the eye pressure. If these measures fail, laser procedures are an alternative to surgery. If the laser technique fails, surgery is necessary.

A procedure called *laser photomydraisis* opens the pupil of

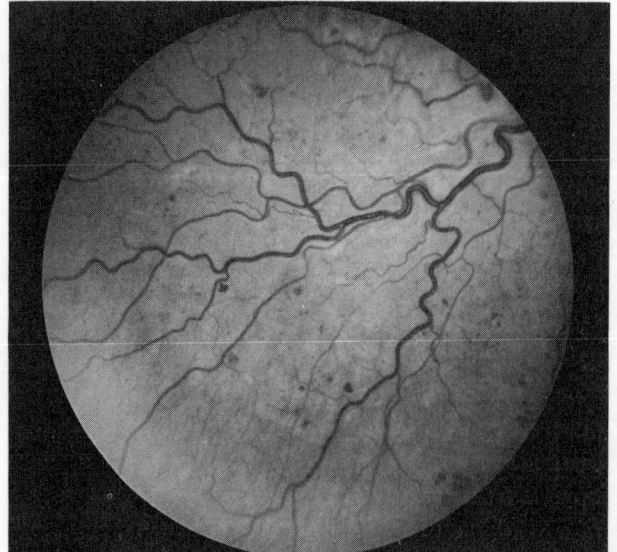

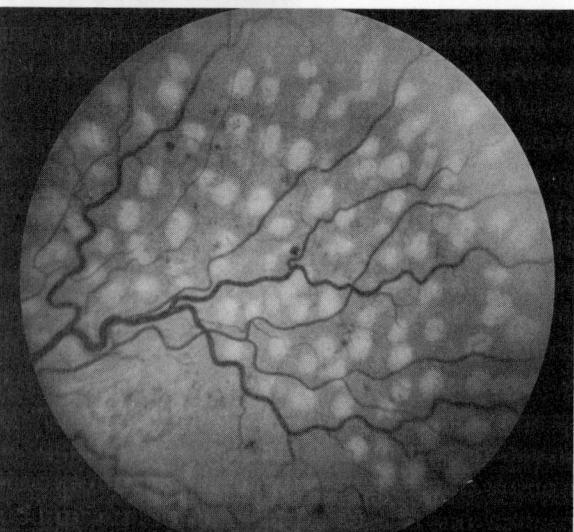

(Photographs by Dwayne Rolf, ophthalmic technician, courtesy of Drs. M. Coleman Driver and Lyle D. Koen, Austin, Texas)

These photographs illustrate the use of panretinal photocoagulation by an argon laser to treat a form of proliferative diabetic retinopathy. The neovascularization is within the optic disk, where direct application of the laser is not possible. The optic disk is the region where the nerves to the brain join the retina. In this case the laser is used to scatter-burn the region of the retina that is well outside the optic disk. The first photograph, before treatment, shows the region to be treated. The macula, the central part of the retina responsible for the central, highest-resolution, and most frequently used part of vision, is not shown but is just to the *bottom left* of the circular portion of retina visible in the photograph. In the second photograph, taken just after treatment, the lighter spots are the laser burns. This form of treatment acts indirectly to reduce neovascularization in and around the optic disk.

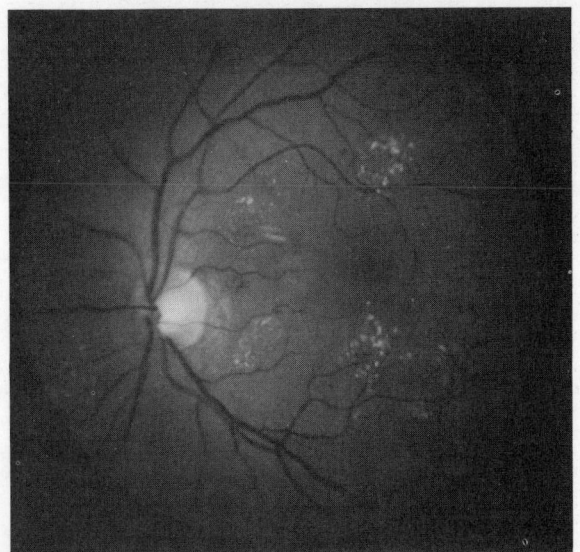

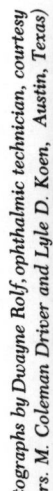
(Photographs by Dwayne Rolf, ophthalmic technician, courtesy of Drs. M. Coleman Driver and Lyle D. Koen, Austin, Texas)

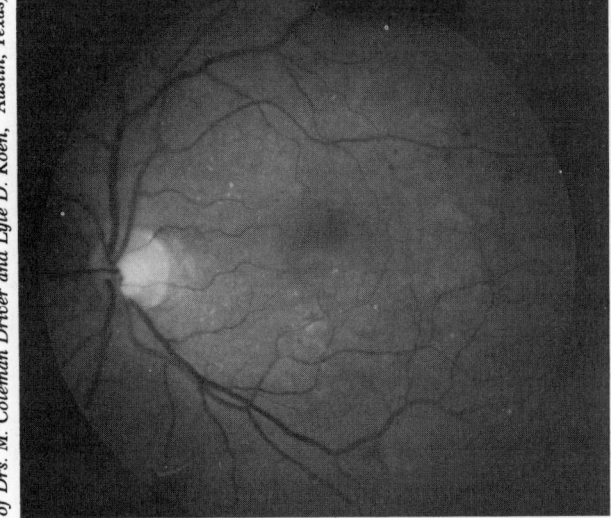

These photographs show argon laser treatment of diabetic maculopathy, which is neovascularization and presence of exudates in the macula. The macula is the circular area encompassed by the two most prominent arcs of blood vessels. The central part of the macula, the fovea, is the dark area just to the *left of center* in each photograph. This is an example of local photocoagulation. In the first photograph, before treatment, the neovascularization is not readily visible, but the white exudates are easily visible. In the second photograph, some time after treatment, the exudates have been absorbed. With careful examination of the photograph, laser-induced scarring in the region of the fovea can be seen. The laser spots are very small and slightly lighter in color than the retinal background.

the eye to counteract the action of drugs called *miotics* that are used to treat glaucoma. Burns placed on the iris cause it to recede and release the sphincter muscle that holds it closed. In *laser iridotomy,* a hole is burned through the iris so that the aqueous solution can drain into the anterior chamber through the trabecular meshwork. This releases excess pressure on the eye. *Laser trabeculoplasty* also assists drainage of the aqueous humor, but in this case, burns are placed directly in the trabecular meshwork. These burns reduce the trabecular ring diameter enough to reopen the spaces through which the fluid must drain.

These procedures are usually done with an argon-ion laser operating at 250 to 600 milliwatts in power. The laser-focused spot size varies from 100 to 500 micrometers. The surgeon varies the exposure time to achieve the desired results. A typical exposure time is 0.2 second. Because use of the laser permits intervention without the necessity of physically penetrating the eye with surgical instruments, most of these procedures can be done in the doctor's office with local anesthetics.

A third application of lasers involves complications that can occur after cataract surgery. Following cataract surgery, a membrane known as the posterior capsule can cloud over, disrupting sight. The mode-locked or Q-switched Nd:YAG laser is used to quickly and painlessly tear open the membrane, restoring vision. This process, referred to as *photodisruption of the posterior capsule,* can be done with artificial lens implants in place.

Since excimer lasers can make precisely controlled cuts in delicate tissues, they are excellent candidates for performing *radial keratotomy,* a procedure that can be used to correct myopia, or nearsightedness. Precise radial incisions in the cornea are used to change the shape of the cornea, alleviating both nearsightedness and astigmatism. Recent research has shown that 4 pulses of 193-nanometer radiation from an argon fluoride excimer laser with energies of 250 millijoules each and durations of 14 nanoseconds removes one micrometer of cornea, thus providing excellent depth control. The incision can be as narrow as 20 micrometers. In some cases, this treatment can improve a person's vision.

Lasers also provide new and better diagnostic tools for the ophthalmologist. In an examination, it is difficult to get a good clear view of the retina. A device called the scanning laser ophthalmoscope (SLO) uses very-low-power argon, krypton, and helium-neon laser beams to scan the retina. The beams are scanned by acousto-optic deflectors and then enter the eye. The beam power is only about 40 microwatts, very small compared to conventional ophthalmoscopes. Reflections from the retina are received by a very sensitive light detector called a photomultiplier tube and from there are sent to a television monitor. The result is a clear, bright picture of the retina that the ophthalmologist can use to diagnose the patient's condition.

CORONARY AND ARTERIAL MEDICINE

One of the most important medical applications of lasers is the removal of deposits that occur in the heart and arteries. These are calcified and fatty deposits called plaque. Millions of Americans die or are disabled each year from cardiovascular disease caused by blockage of the arteries, blockage of the heart, or damage to heart valves. The conventional solution to blockage of arteries is the coronary bypass operation, and the conventional solution to calcified valves is their replacement by artificial heart valves. These are serious and expensive procedures requiring long recovery times. It may soon be possible to remove arterial blockages, remove deposits from the heart, and to clean deposits from heart valves without major surgery. The basic concepts for doing this are in place, and laboratory investigations are being done to perfect the techniques and to solve technical problems. Until it is possible to remove plaque without major surgery, these techniques are aiding doctors working with the surgical methods.

The concept of laser treatment of arterial deposits is to introduce a catheter into the affected heart or artery. This can be done by surgery or by moving a catheter through the veins of the arm or leg into the heart area. The catheter contains two fiber-optic cables that can carry the laser pulses and permit the surgeon to see the results as the process is carried out.

Once the catheter is in the blocked region and correctly positioned, the surgeon begins firing the laser to vaporize the plaque. The vapors are carried away by another tube in the catheter. The surgeon keeps firing the laser until the area is sufficiently clear. Then the catheter is removed, and the process is complete. With this procedure, there is possible damage to the arterial walls. Perforating the arterial wall is a serious problem that would require surgical intervention. This problem must be solved before the technique can be used.

The process of plaque removal by lasers is called *laser angioplasty.* A likely candidate for laser angioplasty is the CO_2 laser because of its absorption by plaque material. However, fiber-optic cables for long-wavelength radiation have just been developed; they are not perfected or readily available. Argon is not a good candidate, because, while the plaque does not absorb it well, the arterial wall does. Pulsed Nd:YAG and excimer lasers are excellent candidates for this process. Excimer lasers appear to be the best choice, provided they can be perfected for surgical use. The same advantages and disadvantages discussed earlier in connection with excimer lasers apply also to their use in laser angioplasty.

ONCOLOGY—PHOTORADIATION THERAPY OF TUMORS

Procedures using lasers and other light sources are being used to selectively destroy cancerous tumors; they have proven very effective and may eventually replace other forms of tumor therapy. With lasers there is no radiation or thermal damage to cause scars or other undesirable side effects. The treatment can be repeated without risk. The laser method works faster and with less trauma. One treatment can kill tumors that would otherwise require weeks of ionizing radiation with all its side effects.

The treatment requires the patient to take a substance derived from cow's blood called hematoporphyrin derivative (HpD). This substance makes the cells of the body photosensitive. The HpD moves out of normal tissues and stays longer in cancerous tissues. When the concentration is maximum in the cancerous tissues and tumors, the region known to be cancerous is exposed to moderately intense visible light at about 100 milliwatts per square centimeter at wavelengths of from 620 to 640 nanometers. The normal tissue is not harmed, but the HpD in the cancerous cells undergoes a photochemical reaction, releasing a form of oxygen that destroys the cancer cells.

Now that this form of therapy is proven, it is being used in conjunction with conventional surgery to help assure that all the cancerous tissue is either removed or destroyed. It can also be used in regions of the body where surgery would be dangerous or fatal.

NEUROSURGERY

The use of the laser in the treatment of brain disorders has been called a revolution in neurosurgery. So far the CO_2 laser has been the most common and most useful laser for neurosurgery. It has the advantages of good absorption, low scatter, and precise controllability. However, in view of the advantages of pulsed Nd:YAG lasers and excimer lasers, this situation could soon change. There is no substitute for the focused beam of energy from a laser to incise or remove tissue adjacent to sensitive neural or highly vascular structures.

Lasers have been very successful for the removal of benign tumors, yielding better results than conventional techniques. Both CO_2 and Nd:YAG have been used for this purpose. With malignant tumors, little or no advantages over conventional treatment have been observed, since adjacent areas have usually been affected by the malignancy. The primary advantage of the removal of brain tumors by laser is the minimization and elimination of trauma to surrounding tissues. The laser is invaluable for selective removal of neural tissue for relief from chronic pain.

The major problem with the application of lasers to neurosurgery is the same as we have mentioned for other medical applications: thermal damage to nearby tissues. With increased perfection and application of pulsed Nd:YAG and excimer lasers, this problem may be completely eliminated. These lasers also have the advantage of easier delivery by fiber-optic cables and focusability to smaller spot sizes for greater surgical precision. Because of the requirement for precision, neurosurgery may be able to take greater advantage of laser technology than other areas of medicine.

GYNECOLOGY

The CO_2 laser is being used in gynecology to remove growths from the cervix that if left can become cancerous. The procedure is called *laser vaporization conization.* Use of the laser results in much less loss of blood than with conventional techniques. The laser method is the most efficient available, and recovery is more rapid since smaller amounts of healthy tissue are damaged during the procedure. Perhaps most important, studies have shown that the laser process does not interfere with fertility. The cervix must remain open and undamaged if conception and childbirth are to occur without complications; the laser method does not interfere with either.

Large numbers of women who suffer from a condition called *menorrhagia* menstruate continuously; a hormone imbalance constantly sends messages to the uterus to shed its inner lining. In the past this condition has been treated by administering hormones or by hysterectomy. Many women cannot take the hormones, and hysterectomy is major surgery. Doctors have developed a technique using a continuous Nd:YAG laser to vaporize the lining, removing it from the uterus. This permanently ends the menstrual cycle. The technique uses 50 watts of power fed into the uterus by a large-core optical fiber. The laser beam is transmitted through a saline solution used to fill the uterus during the procedure, and the beam vaporizes the tissues of the uterine lining.

In gynecology, as in other areas of medicine, laser development is resulting in even better, more efficient procedures. There is a strong possibility that the ideal laser for gynecology will be an excimer laser to nonthermally ablate the tissue,

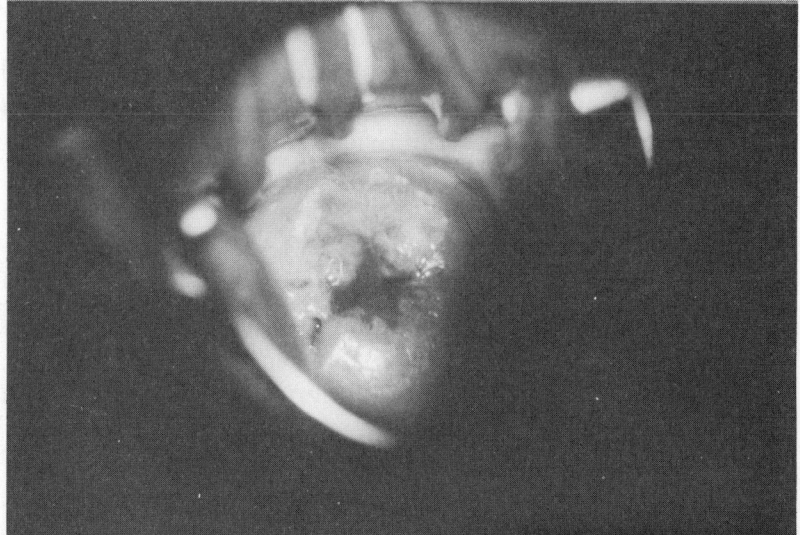

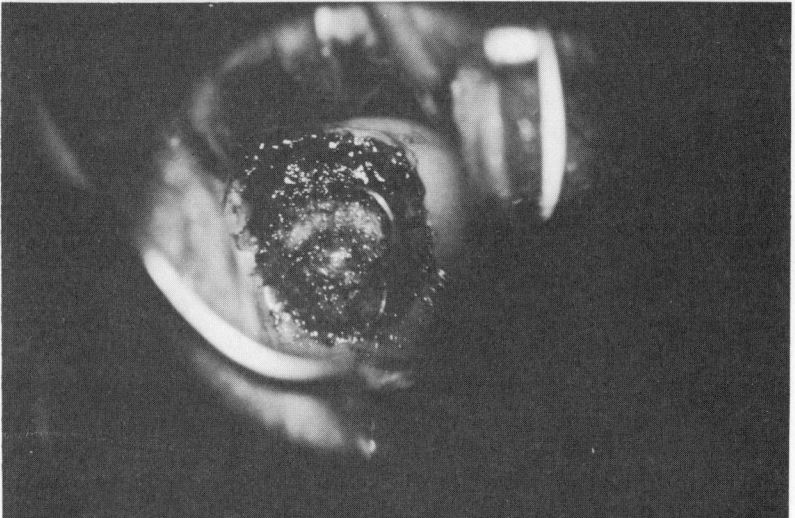

(Photographs by Ms. Debbie Roush, R.N., courtesy of Dr. Harold W. Brumley, Austin, Texas)

These photographs show a cervix before and immediately after laser vaporization conization. The light areas surrounding the opening in the center of the cervix are the growths that may become cancerous if not removed. After removal, the area is charred from the laser burns. The cervix heals rapidly and soon will look normal even though there may be some bleeding and discharge. Cauterization by the laser reduces the chance of infection. With this process, fertilization is not impaired.

combined with output from an argon laser for coagulation of blood vessels.

Many other gynecological problems are being treated with laser systems. These include removal of growths from the vulva and vagina, and removal of blockages at the opening of the oviduct that can interfere with fertility.

PODIATRY

Podiatrists are using CO_2 lasers to assist in treatment of warts, fungus nails, ingrown toenails, nerve tumors, scar tissue, root calluses or clogged pores, hammertoes, and spider veins (clusters of veins on the ankle). Removal of chronic warts with the laser has resulted in an 85- to 90-percent success rate, as reported by one doctor, compared to a 75- to 85-percent rate with existing techniques. The CO_2 laser vaporizes the warts and cauterizes the wound. Even though an open wound results, it tends not to bleed or get infected. Postoperative pain is reduced, and the area tends to heal quickly.

Ingrown nails and fungus nails are treated by vaporizing the corners and roots of the nail rather than removing it. This procedure prevents the nail from growing inward again, and involves less bleeding and pain than traditional treatments. Some treatments that might otherwise require hospitalization can now be done in the doctor's office.

Tumors, scar tissues, and deep root calluses or clogged pores can be cleanly vaporized with little pain or bleeding. Not all medical foot problems can be treated by lasers, nor are some laser treatments better than conventional treatments, but laser treatment is superior for some foot problems because it reduces pain, the chance of infection, and bleeding, while speeding recovery time.

DERMATOLOGY

Some of Star Trek's incredible medical capabilities may be just around the corner. Does it seem like pure science fiction when Dr. McCoy picks up a laser and miraculously runs it along a severe wound, closing and healing it? It's not so preposterous, it turns out. Lasers are being used to weld skin together and to stimulate the healing of chronic human skin ulcers. These procedures do not happen instantaneously, but they have very real possibilities. Using powers lower than those required for surgery, CO_2 lasers can convert the proteins in skin and other body tissues from solid to liquid without burning the skin. As the skin returns to solid form, it stays together, closing the wound and preventing infection. Additional healing remains to be done, but the worst is over. This process eliminates the need for sutures (stitches) and reduces scarring.

It is also possible that laser light can affect healing rates, by stimulating collagen synthesis, which accelerates healing. Polarized light at 630 nanometers has been effective against chronic skin ulcers that fail to respond to conventional methods of treatment. Researchers have shown that the polarized light increases cellular defense mechanisms and mechanisms that remove bacteria from the wound that prevent healing. Other, more complex defense mechanisms are also stimulated. Experiments in collagen synthesis show that low-energy lasers can be used to treat skin sores and enhance surgical wound healing. These mechanisms depend on the fact that the light is polarized. The exact mechanism that permits polarized laser light to promote wound healing is not understood, but it is related to the interaction of the light and the cell membranes.

DENTISTRY

Laser treatment of dentin (tooth enamel) hardens the surface and increases its resistance to acids. The laser causes a crystalline barrier to form that is about 50 to 100 micrometers deep. This is a thermal treatment effect, similar to industrial laser surface treatment, and does not involve cutting or removal of material. This process is very valuable for treating and preventing tooth decay and for treating root infection and other infectious complications. When treating tooth decay, if the decay is deep, the laser can stop bleeding in the pulp and can help sterilize and heal the surrounding dentin. Laser treatments help avoid complications and hasten recovery times.

The coagulation properties of laser treatment are also valuable for surgery of the gums and mucous membranes of the mouth. The laser is used to remove tumors and other irregularities with less trauma (less damage to other tissues). Laser procedures are more sterile, and they aid the healing process.

The CO_2 laser is the best choice for dentin treatment. In oral surgery, just as in gynecology, treatments involving cutting or removing tissue may soon be improved by the use of excimer and argon lasers.

OTHER MEDICAL APPLICATIONS

Diode laser treatment of persons suffering from arthritis is effective in reducing pain, stiffness, and other discomforts. The exact reason for these effects is not fully understood. It appears that the light is capable of inducing chemical changes that relieve the effects of inflammation. The process could be of great value to people who cannot take anti-inflammatory drugs. One theory is that the beneficial effect is the result of increasing collagen synthesis, the same effect that may accelerate wound healing.

There is evidence that the benefits gained from acupuncture can also be gained from laser acupuncture, in which the laser beam simply substitutes for the usual needle, making the process simpler by alleviating the need for material penetration into the skin. Benefits from laser acupuncture have been reported by the Chinese, who used helium-cadmium and helium-neon lasers. In general, these benefits have not been substantiated by the medical community. The argon laser may be a good candidate for laser acupuncture because of its deep penetration of tissues.

Applications in Communications

Lasers are used in two major ways for communications. The first is through the use of optical fibers, and the second is by direct transmission through the atmosphere or through space. Communication by optical fibers is a rapidly growing area and is a major application of diode lasers. Technical developments have been so rapid that optical fiber systems now offer the capability of sending more information through a cable that is smaller and cheaper than conventional wire lines. Besides being capable of carrying more information, optical fibers are not subject to electrical interference as are conventional lines. They are also more difficult to listen in on, an important feature for the military, because there is no external emission of the signal as in electronic systems. The day is near when one fiber-optic cable a few millimeters in diameter will be capable of carrying one million phone calls.

Laser-based systems are also used to communicate by direct transmission through the atmosphere. These are usually employed for short-range data or voice links, such as communication between buildings. They are subject to atmospheric conditions and weather and therefore are not as reliable as fiber-optic systems. The military is developing communications systems for submarines using lasers that emit in the blue-green region of the spectrum, which is where seawater has its best transmission. Someday soon, ground- or space-based lasers using mirrors or satellites will transmit laser beams carrying voice and data communications to submarines under the sea.

FIBER OPTICS

The development of optical fibers began in the late 1960s. The major initial problem to be overcome was *attenuation*, the absorption of the signal as it passes through the fiber. Although today that problem has been solved, fiber-optic systems are limited by other factors. Laboratory demonstrations now show that it is possible to send optical signals through a fiber a distance of 161 kilometers (100 miles) without amplifying the signal. Fiber-optic voice and data communications systems are being installed by telephone and communications companies at an incredible pace worldwide, but especially fast in the United States, Great Britain, France, West Germany, and Japan. Many are being used for local area networks that relay telephone conversations between regional stations and nearby cities. AT&T is planning a fiber-optic network that will span the entire United States by 1990.

The operation of a fiber for carrying optical signals over long distances is based on total internal reflection, discussed in chapter 4. The optical fiber consists of a long, narrow central core of optical material (various types of glass) surrounded by a thin layer of a different type of optical material. The outer layer is referred to as the cladding. A diagram showing the components of an optical fiber is given in figure 7.4–a. The core material is more dense, having a higher index of refrac-

Figure 7.4. Schematic Diagram of a Fiber-optic Cable and Communication System

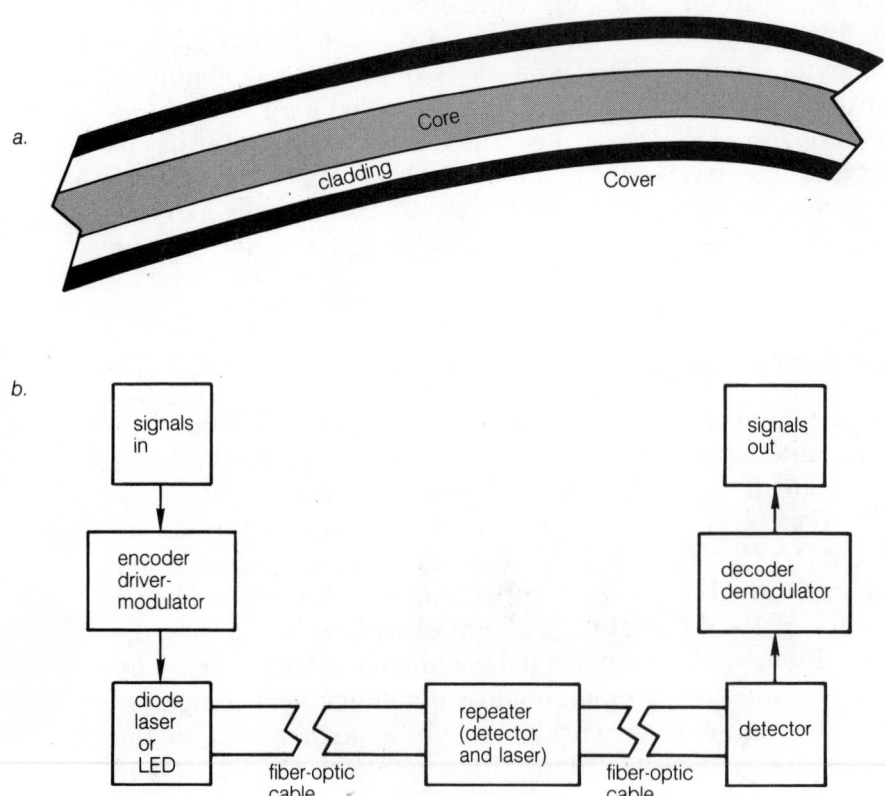

tion than the cladding, a less dense material. When an optical signal from a light-emitting diode (LED) or diode laser is introduced at one end of the fiber, it will tend to stay in the core because of total internal reflection and therefore will be transmitted to the other end, where it can be detected and put to use.

A simplified schematic diagram of a fiber-optic communications system is shown in figure 7.4–b. Incoming signals, which may consist of telephone conversations, computer data channels, or television broadcasts, are multiplexed together (electronically mixed into one signal) and sent to a driver modulator that powers and modulates an LED or a diode laser. The laser light is coupled into a fiber-optic cable. If the distance to be traveled is very long and the signal becomes too weak to be detected at its destination, a repeater is needed. The repeater consists of a detector that senses the signal, followed by an amplifier and another laser that places the optical signal into the next link of fiber-optic cable. At the destination, a detector senses the optical signal and converts it into an electronic signal that is decoded to separate the original signals and make them available for use.

Four major factors determine the capabilities of fiber-optic communications systems: the capability of the LED or diode laser, the losses or attenuation of the signal in the fiber, the bandwidth of the fiber, and the capability of the detector. The bandwidth is determined by *dispersion* in the fiber, which causes the pulses of light to spread.

Both LEDs and diode lasers are used to drive fiber-optic systems. The LED is similar to a diode laser but without the optical resonator. It emits light from spontaneous recombination of electrons and holes. LED emission is not very collimated and has a wide linewidth, from 40 to 100 nanometers. The response of LEDs to modulation is much slower than that of diode lasers, and LEDs are useful for lower-data-rate, less expensive systems. They are easier to operate, last longer, and are less sensitive to changes in temperature. LEDs and diode lasers operate most efficiently when they are designed to emit from 800 to 900 nanometer radiation in the infrared region of the spectrum.

Diode lasers, with their narrower beams and narrow linewidths, have faster modulation response times and are more efficiently coupled into an optical fiber than are LEDs. Coupling loss occurs as the light is fed from the laser into the fiber. Diode lasers are more suitable for long-distance, high-data-rate systems. They can be operated single mode or multimode. In this case, single mode or multimode refers to single or multiple transverse modes. Single-mode diode lasers are more suitable for single-mode optical fibers. Fibers are either single mode or multimode, like lasers. Single-mode fibers are smaller in diameter (about 10 micrometers), to keep down the occurrence of higher-order modes, and are capable of carrying higher data rates over longer distances than multimode fibers (typically 50 micrometers in diameter). Multimode fibers are used with multimode diode lasers and LEDs.

Initially, the major limitation on fiber-optic systems was attenuation of the signal by absorption in the optical material and losses through the cladding. As a result of extensive developments in materials and high-purity techniques for making optical fibers, attenuation is no longer the major problem. Optical fibers transmit best for wavelengths at 1550 nanometers, where they have reached a theoretical minimum of attenuation. The best fibers so far developed lose only about 6 percent of the signal in a kilometer. At 850 nanometers, typical fiber losses are 45 percent per kilometer, and in fibers optimized for transmission at 1300 nanometers, where dispersion is at a minimum, the losses are less than 20 percent per kilometer.

Another limiting factor on optical fiber performance is dispersion. Dispersion occurs because different wavelengths travel at slightly different velocities in the fibers, causing the pulses from the modulated diode laser to spread out as they travel along the fiber. If they spread out too much, the detection system cannot properly interpret the incoming signal. Fibers have lowest dispersion for optical signals at 1300 nanometers. Fibers are now so low in loss that they are said to be *dispersion limited*, which means dispersion is now the factor limiting their performance. Current research efforts are concentrating on developing low-loss fibers at 1300 nanometers and low-dispersion fibers at 1550 nanometers, so that both attenuation and dispersion can be minimized at the same wavelength. And a major effort is underway to develop improved diode lasers and detectors that will operate at these wavelengths as well as they do at 800 to 900 nanometers.

Even though laboratory work has demonstrated the ability to send 400 megabits per second over 161 kilometers without a repeater, more practical and reliable systems that are ready for cost-effective installation are capable of sending only 140 megabits of information per second over 36 kilometers without repeaters. Large-scale fiber-optic communications systems are being planned to supplement and replace existing data and telephone networks, but many plans—especially those for underwater systems where maintenance costs can be very high—are waiting for new laboratory developments so that these systems will be more economical and require less maintenance. A minimal number of very-high-reliability repeaters is essential for the operation of a successful system.

Fiber-optic systems are installed and in use in many areas of the United States, France, West Germany, and Japan. Other fiber-optic communications systems are being planned for various parts of the world. A transatlantic link between the United States and England is scheduled for completion within this decade. Optical signals in fiber-optic cables may soon carry more information than electrical signals in wire cables.

ATMOSPHERIC OPTICAL COMMUNICATIONS

Atmospheric optical communications systems consist of a modulator and laser-transmitter unit that sends communications through the air to a detector-receiver and demodulator unit. These systems can be used for voice, television, and computer data transmissions. They are generally limited in use to less than 20 kilometers. They are less costly, easier to install, require no FCC license, and perform better than microwave communication links. These systems are used by hotels to relay television programming from building to building, by television stations to relay broadcasts to recording equipment, by businesses to provide voice and data relays, and by the military for secure ground-to-air, air-to-air, satellite-to-ground, and satellite-to-air communications. The military systems are much more complex than the commercial ones since the communicating elements are moving with respect to each other; they require acquisition and tracking systems to initiate and maintain the link.

Early designs of atmospheric communications systems involved the use of gas and solid-state lasers. Now diode lasers are used in nearly all short-range systems. The diode laser's compactness, simplicity, reliability, and low-operating-power requirements make it singularly desirable for this application. For this reason most of these systems operate in the infrared region, from 800 to 1300 nanometers. The receivers contain optical filters with very narrow bands, to block out unwanted optical interference such as sunlight, scattered daylight, street lights, automobile lights, and the like. A large number of nonmilitary optical communications systems are commercially available.

The military is particularly interested in optical communications systems because of their so-called low probability of intercept. These systems are excellent for secure communications. The narrow optical beam cannot be intercepted unless the interception is made almost directly between the transmitter and the receiver. Also, the optical system does not inadvertently radiate signals into the surrounding area, as radio and microwave links do.

The United States Air Force is developing an optical system for aircraft-to-aircraft communications in tactical air combat, based on pulsed gallium arsenide diode laser arrays that develop 100 watts at peak power. Each aircraft has a laser transmitter and detector-receiver operating in conjunction with an acquisition-and-tracking system. The transmitting direction must be rapidly and precisely controlled. The major problem to overcome is finding the other aircraft's signal so that it can be tracked.

The United States Navy is developing an elaborate optical system for communicating with submarines. The system takes advantage of water's strong capability to transmit blue-green light. The system is conceptualized to consist of a very-high-power laser operating from a satellite, which would receive signals from other military communications satellites and beam down vital operational information to submarines that are submerged up to 100 meters (328 feet). The laser being developed for the purpose is a frequency-shifted xenon chloride excimer laser. The developmental laser is capable of generating 4-joule pulses of 308-nanometer radiation in the ultraviolet ragion of the spectrum, with an average power of 400 watts. The output light is frequency-shifted to the blue region at 459 nanometers using a nonlinear optical process called Raman frequency conversion that occurs in lead vapors. This wavelength is perfectly matched to special atomic cesium filters that will be mounted on the submarine to highly discriminate against sunlight, picking out only the laser light at the correct wavelength. The filter must very effectively block out all unwanted light, and the detector must be very sensitive, since not much light will remain by the time it reaches the ocean depths.

Lasers are making possible a new era of communications by optical signals that have higher capacity and are more efficient while being less expensive.

Military Applications

The military has found the laser to be remarkably useful. It continues to expand our defense capabilities every day through the use of lasers and their associated optical and electronic systems.

The army likes the idea of communicating with lasers since they are difficult for the enemy to intercept, thereby ensuring the secrecy of critical communications. Numerous "low probability of intercept" optical systems are being used by the army for battlefield communications.

The air force has developed laser-guided bombs, which have optical systems that steer toward light reflected from a target, requiring only one or two bombs, rather than hundreds, to destroy a target. The illumination of a target with a laser is called *laser designation*. Laser designators can be airborne in fighters, helicopters, and remotely piloted vehicles, or ground-based with troops. Once the target is illuminated, warheads from artillery or missiles that can see only the laser wavelength can correct their own course and be almost assured of hitting it.

The army uses diode lasers to build battlefield-simulation systems for training purposes that alleviate the need to use real or dummy ammunition. Firing a weapon is represented by firing a diode laser through an optical system. Hitting battlefield targets is simulated by detectors on vehicles and troops that are tuned to receive only the wavelength of the diode laser that represents a weapon.

The air force is very interested in high-power lasers that can be used to shoot down aircraft, missiles, or satellites. For this purpose, they have developed gas-dynamic CO_2 lasers and chemical lasers. A powerful 400-kilowatt gas-dynamic CO_2 laser has been flown in a specially equipped aircraft and used to disable air-to-air missiles. Most development of lasers for high-power military applications is now concentrated on chemical lasers. Attention is also turning toward the development of high-power lasers at shorter wavelengths. Short-wavelength lasers can be collimated over longer distances, and therefore they can deliver more energy onto a target.

The army is developing ground-based laser systems to disrupt the enemy's electro-optical sensors. These sensors perform observations, acquisition and tracking, ranging, targeting, and weapons control. The army's laser system is designed as an electro-optical countermeasure and is tank mounted. It can also be used to disable or even blind enemy troops.

Fiber-optic systems are being used in aircraft design to send information from one part of an aircraft to another. They provide clean, clear transmission of critical data among navigation, location, targeting, and control systems. They are also used to send critical information to the pilot. Fiber-optic communication is not subject to electrical interference and cannot be picked up from the outside by enemy sensors.

Holographic techniques can be used in conjunction with an aircraft's indicator systems to project information into the pilot's view without requiring him to look down at instruments. These *heads up display systems* permit the pilot to see vital information without distracting him from important flight tasks.

The ring laser gyroscope, discussed under engineering applications, is the accurate and reliable heart of the navigation and control systems for many missiles and military aircraft.

The military is finding many other uses for lasers that are too numerous to cover here. Let us take a closer look at the topics of laser designation, tactical weapons, strategic weapons, and battlefield simulations.

LASER DESIGNATION

The army and air force have developed many laser designation systems, some carried by ground troops, some mounted on battlefield vehicles, and others flying in aircraft or helicopters. All of these enable military personnel to locate targets and place laser light on them long enough for an optically guided warhead or missile to attack them. Some of the more sophisticated designation systems are capable of automatically tracking a target and keeping the laser in place long enough to permit laser-guided warheads or missiles to arrive. They also have the necessary electronics to use the laser light reflected from the target to measure range.

The military has several laser-light-seeking munitions. The army's Copperhead warhead is launched from a cannon. The army also has the Hellfire missile. The navy has a laser-guided 5-inch warhead, and the air force has laser-guided bombs. All of these weapons are designed with optical sensors tuned only to the light from laser designators. These optical systems are capable of sensing, during the last few moments of flight, errors between their flight path and the location of their targets. Error messages are sent to aerodynamic controls that make appropriate corrections to the flight path.

A new and versatile system for finding and designating targets without exposing troops to the severe dangers of going near or entering enemy territory has been developed for the army by Lockheed Missiles & Space Company in Austin, Texas. This system consists of a small computer-controlled aircraft,

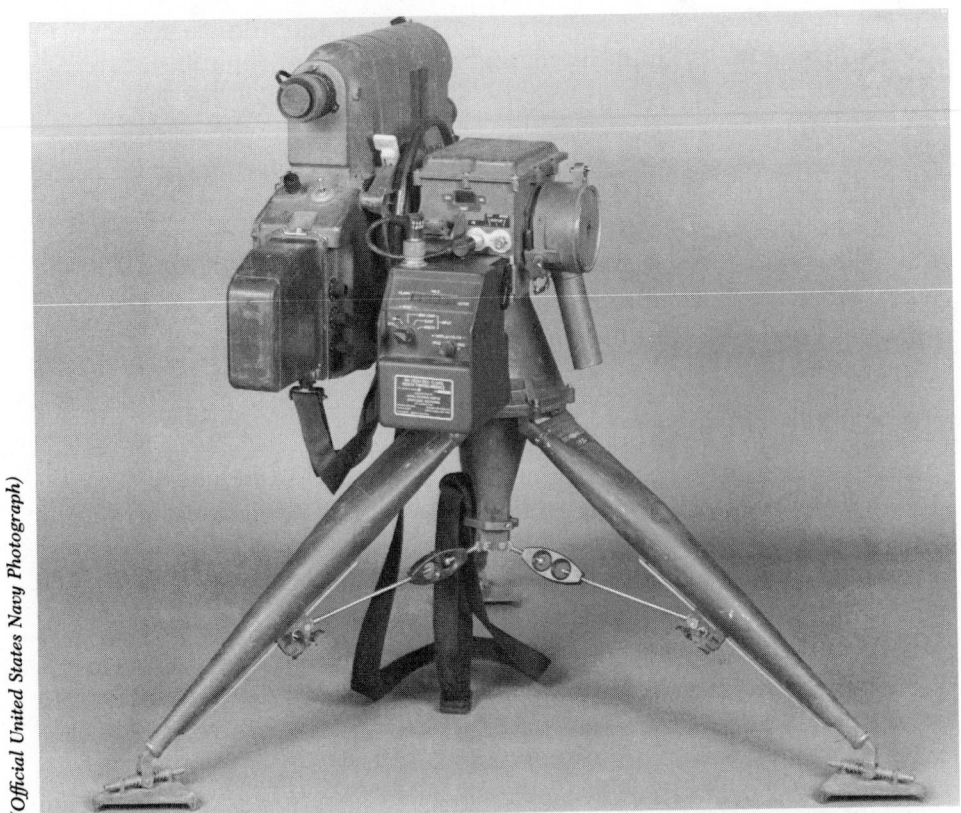

(Official United States Navy Photograph)

This ground-based laser designator and range finder is called a Modular Universal Laser Equipment, or MULE. There are hand-held versions, but this tripod-mounted unit offers greater designation stability.

called the *Remotely Piloted Vehicle,* or RPV, that can be launched by a catapult from a truck in the battlefield. It can fly behind enemy lines, send back daytime television or nighttime infrared pictures, track targets, laser designate chosen targets, and return to the launch point to be recovered in a net, all under the control of operators located in a ground-control station outside of enemy territory.

The heart of the RPV is the *air vehicle,* a small aircraft 4 meters (13 feet) in wing span, 2.1 meters (7 feet) long, and 117.25 kilograms (259 pounds) in weight. It is driven by a small, two-cycle, rear-mounted engine driving a pusher propeller. It contains its own on-board navigation system, air-to-ground data communications system, and autopilot computer. It can be equipped with either a daylight television camera and tracker built by Westinghouse Electric Corporation, or a night-vision forward-looking infrared image camera and tracker built by Ford Aerospace Communications Corporation. Both of these camera-and-tracking systems are controlled by a ground operator. On operator command, they can lock onto and track targets. They also contain a laser transmitter and a laser receiver that not only designates the target but measures its range. The camera system tracks and automatically keeps the laser on the target until the laser-guided warhead or missile has arrived. After the target has been attacked, the operator can assess the effectiveness of the attack and report the results.

The laser transmitter is a small, compact neodymium (Nd:YAG) solid-state laser manufactured by the Laser Systems Division of Litton Systems, Inc., of Orlando, Florida. The

A Remotely Piloted Vehicle is shown in flight. The microcomputer-controlled television and laser designation system views out from the turret located underneath the air vehicle. The optical system is controlled from the ground and can look in almost any direction below the air vehicle.

(Photograph courtesy of Lockheed Missiles & Space Company, Inc., Austin Division, Austin, Texas)

infrared output of the laser is not visible to enemy troops unless they have specially equipped goggles. The laser is pumped by a pulsed flash lamp and is modulated with a secret code so that the incoming warhead or missile will know which target to attack, in case there is more than one target being designated. The laser receiver looks for reflections of the laser light from the target and with high-speed electronics measures the time it takes the laser light to get to the target and return. This determines the target's range. With knowledge of the target range and the position and altitude of the aircraft, it is possible to determine the position of the target and report it to weapons fire control stations.

While the air vehicle is in flight, controllers on the ground are required only to tell the aircraft where to go and how high to fly. The on-board computer takes care of the rest, controlling the engine speed and controlling the flight by adjusting the elevons. When its mission is complete, the RPV aircraft returns on command from the ground and automatically flies into a large net suspended above a truck. It can be immediately refueled and flown again. The RPV system provides a means for surveying a battlefield situation, locating and following targets, and aiding, with a laser, the attack of targets without endangering army troops.

TACTICAL LASER WEAPONS

Although most tactical military laser applications involve target sensing, ranging, tracking, designation, and communications, weapons-oriented systems are being developed by the army and air force.

The army's project Stingray is an optical and electro-optical countermeasure system. It is being designed to disrupt the systems that an enemy would use to acquire and track targets. The Stingray system, based on a 100-millijoule pulsed solid-state laser, is to be mounted on a tank. The laser system could disrupt or damage electro-optical systems up to 8 kilometers away and could inflict eye injury at even greater distances.

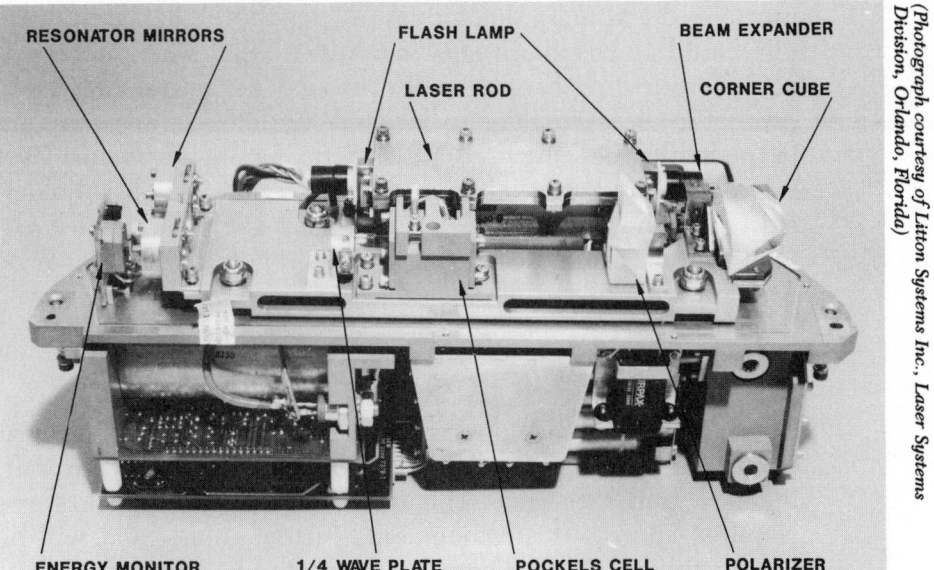

RESONATOR MIRRORS   FLASH LAMP   BEAM EXPANDER

LASER ROD   CORNER CUBE

(Photograph courtesy of Litton Systems Inc., Laser Systems Division, Orlando, Florida)

ENERGY MONITOR   1/4 WAVE PLATE   POCKELS CELL   POLARIZER

The RPV Laser Transmitter is shown with its associated electronics. The laser equipment is *above* the center horizontal plate, and the electronics are *below*. The flash lamp and laser rod are in the housing covered with a flat plate. The large electrical connectors at each end of the housing are for the flash lamp. Both resonator mirrors are located on the same assembly. The large optical component to the *right* is a corner cube designed to fold the beam across the front of the assembly and through several optical components. These are, from *left* to *right,* the polarizing prisms, an electro-optic Pockels cell, and a quarter wave plate. The device to the right of the front resonator mirror monitors the presence of laser output. When the Pockels cell is turned on by the electronics, the polarization of the laser light changes, permitting it to emerge from the prism assembly, sending it over the top of the Pockels cell. The Pockels cell also modulates the designation code into the laser beam.

The air force is developing a system similar to Stingray called Coronet Prince. It is designed for pod mounting underneath an aircraft. Requiring a more powerful laser than Stingray, this system could be used to damage electro-optical display systems and forward-looking infrared (night vision) imagers on board enemy aircraft. It could also be used to confuse the infrared guidance systems of heat-seeking missiles and to injure the eyes of enemy pilots.

While not admitting to the development of lasers for the

purpose, the military recognizes the possibility that laser weapons could be used directly against troops. The military is actively pursuing countermeasures such as special goggles to protect soldiers from enemy lasers. Whether laser weapons are in the field or not, the mere possibility of going up against laser systems in the battlefield could have tremendous psychological effects on troops. Let us hope that the laser, which has contributed so significantly to the quality of life, will never be used for such purposes.

STRATEGIC LASER WEAPONS

There is widespread interest in the possible use, in the next century, of high-power lasers to shoot down enemy aircraft, missiles, and satellites. The development of these systems requires enormous advancements in technology and will be extremely expensive. The development of laser-based systems capable of damaging large objects thousands of kilometers away and traveling at supersonic speeds will not only require lasers significantly more powerful than those we have today but will require considerable advances in the technology of the sensors and computer systems required to acquire, track, and identify potential targets, to control the laser, to keep it on target, and to assess target damage.

The basic concept of an antiballistic missile system is to place in orbit many satellites with lasers capable of several megawatts of continuous output. A mirror system, driven by a pointing and tracking system that identifies targets, would direct and focus the laser energy on targets thousands of kilometers distant. The lasers would damage the missiles by detonating their fuel, detonating their warheads, damaging the surface, or disabling the guidance-and-control system. Enemy strategic bombers could also be attacked by the same system.

The development of high-power strategic lasers is still very much in the laboratory. The air force in its Airborne Laser Laboratory tested the use of a 400-kilowatt continuous gas-dynamic CO_2 laser in a large military aircraft. After several failures, the system was eventually able to acquire, track, and disable several air-to-air missiles fired from another aircraft.

The success of this type of system depends not only on the reliability and capability of the laser but on the ability of the acquisition and tracking systems to do their jobs. High-speed, precision electro-optical and servo-controlled mechanical systems to search for targets, track them, and accurately direct the laser energy to the right point on the target will require advances over current technology. As a result of experience with the Airborne Laser Laboratory, the military's emphasis has shifted toward development of chemical lasers for strategic laser weapons.

Chemical lasers are being tested at the High Energy Laser Systems Test Facility at White Sands Missile Range. The latest laser under development operates with continuous output in the mid-infrared range. Although its power level is classified information, the device is probably capable of delivering more than 2 megawatts. The goal of strategic laser weapons development is to develop a laser capable of 5 or more megawatts of continuous output, which is roughly what will be required for a space-based antiballistic missile system. The Defense Advanced Research Projects Agency is currently sponsoring development of a 5-megawatt hydrogen fluoride chemical laser.

Building lasers in the laboratory is one thing, and putting them in space is quite another. To be practical in space, they must be made highly reliable. Their control systems must be totally automated. They must be lightweight and capable of withstanding launch loads and vibrations. They must be provided with both electronic power supplies and supplies of chemicals to run the laser. And they must be designed to operate with the targeting and control systems that will direct their energy to the required place at the right time. Meeting these requirements, as well as those of the sensing and beam-directing systems, will be a challenge far beyond those attempted to date.

Besides the technical requirements for putting large lasers into space, there is the major drawback of their vulnerability. Large-scale space-based systems are nearly impossible to hide and are therefore very difficult to protect. An alternative solution is to keep the lasers on the ground and to put control systems and relay mirrors on satellites in space. To minimize the laser power that would be required, lasers would be placed on mountain tops. The surveillance, target-acquisition-and-tracking, mirror-relay, and control systems would be in a near-earth orbit. There would have to be many satellites so that one would always be available to the laser base. Geosynchronous orbits, by means of which satellites always remain above the same point on earth, are too distant to be used for laser relay. Ground-based systems will require even higher-power lasers than space-based systems to overcome the transmission losses involved in going up through the atmosphere. Detailed systems studies will need to be done to determine whether ground-based or space-based laser weapons are the most feasible and cost effective.

The air force is investigating the possibility of developing laser antisatellite weapons. They could be based on the ground, in aircraft, or in space. A ground-based laser antisatellite weapon is considered by many scientists and engineers to be technically feasible at this time. Although research associated with such a system could contribute significantly toward a technical basis for an antiballistic missile defense system, at the present time there is not a sufficient technical base to begin design and development of such a system. Since such enormous research and technical developments are needed, a major problem in developing these types of systems will be the staggering cost.

BATTLEFIELD SIMULATIONS

Since 1979, the army has made extensive use of realistic battlefield-simulation systems based on the use of diode lasers. In these systems, the firing of a weapon is simulated by the firing of a pulsed diode laser. The diode laser and its associated electronic and optical systems are mounted on the weapons and bore-sighted to the weapon's aim point. Almost any type of weapon, from hand-held M16 rifles to tanks and artillery, can be simulated. The receipt of weapons fire is simulated by equipping troops and vehicles with silicon detectors that receive the incoming optical pulses.

The laser transmissions are code modulated to identify the firing weapon and its type. The transmission is also coded with sequences of words in messages to enable the receiver to determine whether or not a hit or a near miss has occurred. Hits and near misses are recorded, and the receiver is informed that he is being fired upon so that he can take evasive maneuvers. The receiver-detector sends the incoming signal to a microcomputer that decodes the identity of the weapon and uses lethality logic (kill probabilities) to compute whether or not a hit or a kill has taken place. It sends the results to a central location that keeps track of the simulation results.

These systems are able to provide greater realism than any previously developed battlefield-simulation systems. By tailoring system parameters, such as the laser output, beam divergence, detector threshold, numbers of detectors, numbers of words in messages, and types of messages, it is possible to simulate the kill probability involved in the use of various types of weapons. The systems are able to operate in full sunlight with laser output energies that are safe for the eyes of troops.

This has been only a sampling of the many uses of lasers that improve the capability and efficiency of the military. As the demands placed on the military by the world situation increase, it seems likely that the laser will continue to enhance and expand our nation's defensive and offensive capabilities. Hopefully, as the military becomes more efficient and capable, less loss of life will be required to resolve conflicts. Perhaps we are headed toward the day when wars will be fought by robots and computers.

Lasers in Entertainment

The intense collimated beams from visible output lasers are ideally suited for projection of large images over great distances. Scanning mirrors and modulators are used to form the images and to make them move, change color, and fade in and out. Impressive displays are being created by projecting images stored in computers or synthesized from music signals onto clouds, the sides of mountains, and the sides of buildings.

Helium-neon lasers, with their bright red output at 632.8 nanometers, are reliable and inexpensive. Argon-ion gas lasers have powerful outputs of green at 514.5 nanometers and of a deep, richly colored blue at 488.0 nanometers. Krypton gas can be used alone or mixed with an argon laser's gas to provide red output at 647.1 nanometers. The helium-cadmium laser has a powerful output available in the blue region at 442.0 nanometers. These lasers form the basis for a new and rapidly growing industry that applies lasers to create novel and exciting light displays.

Laser-light displays are most popular at rock concerts, where the displays present their own preprogrammed image sequences or are synchronized to form images that respond to the frequency content of the music. Indoor light displays are frequently given at planetariums, which give the display designer a full 360 degrees for projection. Film studios are using lasers and laser control systems not only for special effects in movies but to create exciting displays for visitors. Amusement centers are using laser-based entertainment systems to create unusual animation sequences that are stored in computers linked to the optical system. Fiber optics also play an important role in these systems. Unusual and highly elaborate lighting effects are achieved without inordinate numbers of lasers or lamp sources. Hundreds of fibers can deliver the output of one laser to hundreds of different locations at the same time.

Laser displays are based on controlling the output of one or more lasers using either optical modulators driven by digital computers or analog signals derived from voice or music signals. A sound-and-light synthesizer system modulates and scans a helium-neon laser in response to the harmonic content and intensity of the music. Different color lasers can be mixed to form multicolor displays. Systems like these can be used at concerts, hooked up to stereo systems, or driven by computers.

Laser animation and display systems use acousto-optic or electro-optic modulators to deflect and modulate the intensity

(LASERIUM Photograph courtesy of Laser Images, Inc., Van Nuys, California)

This colorful figure is formed with scanning mirrors using argon and krypton lasers. When seen live, the figure develops, grows, moves, and changes colors. Individual artists can synchronize the evolution of the figure to music. The control is by means of taped analog signals.

of the laser beams. Colors from different lasers are modulated and mixed together optically to form the final image. Rapid-scanning effects are achieved using flat pivoting mirrors or spinning polygonal mirrors, like those used in laser printers. The modulators and deflectors are controlled by signals that come from microcomputers programmed to control the delivery of stored digitized images. These images can be stored on magnetic disks, video disks, or optical-data-storage disks.

The major problem in generating images with lasers is that the images must be redrawn very rapidly or the image fades. Laser scanning cannot be done nearly as fast as the scanning electron beams in television sets. Further, laser projection systems contain no phosphors, like those in the television screen, to momentarily maintain light emission and give the

(Photograph courtesy of LaserMedia, Inc., Los Angeles, California)

Laser beams and fireworks set up a spectacular display of colors at Stone Mountain Amusement Park in Atlanta, Georgia. Animated graphics that are formed from laser beams and controlled by microcomputers are projected on-to the largest screen in the world, the side of the mountain.

image continuity. At the moment the laser leaves a given point on an open screen, the illumination is gone, and it must be retraced very rapidly for the image to appear to have uniform brightness. To generate images over wide angles, scanning mirrors are needed, and these have response times that are sufficient only to generate rather simple images before retracing is required.

As we saw earlier in this chapter, there are enormous problems to overcome before holography can provide a viable form of technology for three-dimensional display. However, other possibilities are emerging, any one of which may eventually lead to three-dimensional movies or television. A new approach to the problem has been developed by engineer Rudinger Hartwig at IBM's Heidelberg Scientific Center. In his concept, a large glass cylinder, which can be viewed from any direction, contains a helical surface that spins about the cylinder's vertical axis. A laser beam directed into the cylinder from the top will reflect off the spinning surface toward the viewer. As the surface spins and the laser beam is reflected along the cylinder's radius, the viewer receives the impression of a three-dimensional image. By synchronizing the laser output with the rotation of the helix, a spot can be made to appear anywhere within the volume of rotation of the cylinder. Although in the early stages of development, this system may be the basis for three-dimensional displays of the future. These displays will be of great value, for example, in air-traffic-control systems and in medical-display systems. In air-traffic-control systems, operators would be able to more rapidly and accurately interpret air-control radar information. In medical-display systems, locations of tumors could be precisely shown, aiding surgery and diagnosis. Perhaps eventually a system of this type will lead to three-dimensional television.

In the long run, the most widespread use of lasers in the entertainment industry will be the video disks and audio disks discussed in connection with optical-data storage. Presently, the video disk is not doing well in competition with video tape cassettes, because the video disks are not erasable. With the inevitable development of an erasable optical-storage material, there will most likely be a resurgence of the video-disk industry, just as the audio-disk industry is strongly emerging today.

Summary

In only twenty-five years, the laser has evolved from a laboratory curiosity, a device looking for an application, to a major element of our technology. Lasers make it possible to know more about the structure of atoms, to capture three-dimensional images of objects, to measure surface contours within fractions of a micrometer, to measure long distances with great accuracy, to store large quantities of information in very small spaces, to process materials with increased accuracy and speed, to communicate more easily and cheaply over long distances, to restore detached retinas, to open blocked arteries, to guide warheads and missiles to their targets, and to enjoy brilliant images moving in response to music. Many applications, such as controlled-fusion energy production and missile defense systems, are still speculative but are based on sound technological projections. These capabilities and many others that we cannot even begin to visualize will no doubt be part of our future. The laser has contributed substantially to our quality of life and will continue to do so beyond our wildest dreams. Few single technical developments, outside of the discovery of electricity and the invention of the transistor, have had more widespread and significant impact on the world than the invention of the laser.

In the last chapter, we will talk about what you can do to advance your knowledge of lasers, and what you can do to beam yourself down a career path that involves lasers.

Education and Careers in Laser Technology

LASERS are finding so many diversified applications in research, engineering, industry, medicine, and business that persons trained in almost any field have a chance of getting involved with lasers. In chapter 7 we presented a few of the more important and most common laser applications. There are many hundreds of other applications that we have not been able to cover in this brief introduction. New and exciting applications for lasers are being found every day, and soon almost every person's life will be enhanced by the services of lasers and systems based on lasers.

The majority of laser-related careers currently available reside in the telecommunications and aerospace industries. The aerospace firms research and develop laser, optical, and electronic technology required to build defense and aerospace equipment. The emphasis in the aerospace industry is on development of specific technical systems and how to meet the requirements of those systems. Technical advancements and new developments are more likely to be made in research laboratories or special research divisions. While most are at colleges and universities, the government funds several major research laboratories, such as Lawrence Livermore, Los Alamos, Sandia, Aerospace Corporation, MITRE, and Battelle. A few very large corporations like AT&T, Lockheed Missiles & Space

Company, Hughes Aircraft, Rockwell International, AVCO, Northrup, and IBM have their own research laboratories.

Telecommunications companies like AT&T, GTE, IBM, and Corning employ large numbers of technical and engineering professionals to research and develop communications systems. Their primary interests are in fiber optics, diode lasers, and solid-state detectors.

Laser and laser-related equipment, accessories manufacturing, and marketing companies play a critical role in the laser field but do not employ as many engineers and technicians as are employed in the aerospace industry. Besides an advanced degree, it is likely that special experience will be required to secure engineering positions with these companies. If you have a strong interest in lasers, these are some of the best careers to shoot for. These companies have the advantage of being specifically dedicated to lasers and laser-system components and accessories. Smaller companies are more likely to offer financial benefits to engineers in the form of stock options or partial ownership. They sometimes even partially or wholly subsidize a higher degree. At a minimum they offer in-house training as a means of further education. In these companies having a master's degree is essential, and a doctorate is desirable. However, special experience or expertise in an area of company interest could be more important than an advanced degree.

In the laser manufacturing industry, it is difficult to pin down the requirements for technicians. Holding a degree from a vocational school or community college is important, but here too experience in an area of special interest to the company could be more important. Many companies consider relevant experience backed by your previous employer's recommendations to be a more reliable indicator of your potential and your quality of work than whether or not you have obtained a particular degree.

The type of involvement you will have with lasers depends on whether you are interested in lasers themselves, the way they work, and how they are developed, or in how lasers are applied to research, industry, medicine, or business. The former is a *research and development* orientation, in which the laser itself is your primary interest. The latter is an *applications* orientation where the application is your primary interest and is more closely related to your training or degree area. Even if you are applications oriented, you will want some laser-related technical grounding to get a feel for wavelengths, power levels, power densities, focusing, the nature of laser-beam interactions with materials, and the design requirements of various types of lasers.

If your primary interest is the laser itself, the most important disciplines to pursue are physics, mathematics, optics, and electrical engineering. A person who wants a career in laser design and development would most likely major in physics or electrical engineering. Most universities now offer courses in either the physics or electrical engineering departments on laser physics and engineering, usually at the junior or senior undergraduate level. These courses teach all of the subjects you have learned about in this book and go into them in much greater depth. In addition, a course in optics, for an understanding of how laser light can be handled is desirable. If you are primarily interested in lasers, a laser engineering course is a must. If you are applications oriented and have a degree in a nonlaser discipline, a college course of this type would be very valuable. Be sure to verify that you can meet the prerequisites, unless you intend to audit the course. The prerequisites would most likely be freshman or sophomore physics and a course in electromagnetic theory. Such a course may go into a level of detail beyond that required by an applications-oriented person, but it will enhance your understanding of what you have learned here.

The physics major with an interest in lasers should concentrate primarily on atomic physics, optics, and engineering mathematics. Courses from an electrical engineering department in electromagnetic theory are also essential. Theoretical and nuclear physics come into play only in the development of the most exotic lasers envisioned today, the X-ray lasers. These lasers have not been discussed here since very little information is available on them. Most activities with X-ray lasers are classified by the government. For a career in laser or laser-related physics, whether it be in design of lasers or laser systems, a master's degree is essential and a doctorate very desirable. Unless you work your way up to a research position, private laboratories, government laboratories, and laser manufacturers would not be likely to hire you into research and design without an advanced degree.

Applications-oriented persons have many options open to them for learning more about lasers and what can be done with them. There are many excellent books that go into more depth than this one does. Some are available at local bookstores, but your best bet is to go to a technical library at a college or university. Besides undergraduate engineering courses, there are special one- or two-week courses given by various universities and professional societies from time to time. The best way to find out about these courses is to belong to a professional physics, electrical engineering, or optical association. Belonging to one of these organizations gets you on mailing lists of persons who are interested in lasers. You will receive solicitations to attend conferences on lasers and laser applications, where you can get the latest information on developments in many different laser-related fields. While covering many different topics, these conferences are usually highly specialized, so expect to hear presentations in depth on narrowly defined topics; they are not designed to present general concepts. They do provide an excellent forum for exchanging information with others.

Another excellent source of information are monthly professional laser magazines. If your work is in a technical area, you may be able to qualify for free subscriptions. These magazines have abundant information on the latest developments in lasers and their applications. Their advertising will familiarize you with the companies that manufacture lasers, laser accessories, and laser-related systems. Following this chapter is a reference section with a short list of books on lasers, laser magazines, professional organizations, and laser conferences.

If you are interested in a career as a laser technologist, many technical schools offer laser technology programs. If your school does not have a laser technology program, the best areas of emphasis would be electronics and optics. Technicians are needed to install, set up, check tolerances, check specifications, operate, trouble shoot, and repair lasers and related equipment. Other types of positions may call for the operating of process equipment; the reading and interpreting of wiring diagrams, blueprints, and schematics; and recording, logging, and maintaining data.

Setting up and working with laser systems requires a thorough knowledge of various types of laboratory electronic devices, including the types of electronics associated with laser systems, laser accessories, and detection systems. Almost all of these contain power supplies, meters, regulation equipment, and signal-processing equipment. Modern commercial lasers consist of many optical and electronic components and their supporting structures. Manufacturers usually provide detailed instructions, but experience may be required to ensure optimal operation and durability, especially for many special types of lasers found in research laboratories. Laser accessories include items like power meters, spectrum analyzers, mode-locking devices, Q-switches, acousto-optic modulators, and electro-optic modulators. All of these require not only proper setup, use, and adjustment of the electronics but also proper positioning and alignment of the optics.

Manufacturers' specifications must be carefully followed when it comes to the use of detection equipment and cameras. Incorrect settings on the electronics can result in faulty or incorrect readings, and excessive incident light power can damage detectors and electronics. For very sensitive detection systems, liquid-nitrogen cooling may be required and additional electronic devices such as lock-in amplifiers may be necessary.

Besides setup, regulation, and proper use of many different types of electronic equipment, laser technologists must be familiar with basic optics, and they need both laboratory and practical experience with positioning, aligning, and tuning of optical devices. Some optical devices common to laser laboratories are spectrometers, attenuators, filters, lenses, reflectors, polarizers, wave retardation plates, prisms, Pockels cells, Kerr cells, waveguides, fiber-optic cables, and fiber-optic connectors.

Potential laser technologists should seek out technical

schools that can provide either full-scale programs in laser technology or courses in both electronics and optics. These courses should include experience with lasers and optical devices. About thirty colleges in the United States offer programs in laser technology, and some even have programs in fiber-optic technology. It usually takes two and a half to three years to complete such a program. Experience gained from hands-on laboratory work is essential, while final expertise will be gained in the work place.

As laser applications spread, the demand for laser and electronic technicians is rapidly increasing. They are needed in research facilities, universities, industry, hospitals, and businesses. In industry, workers are being replaced by skilled technologists who set up, monitor, and repair automation equipment.

Most persons who want to work as laser, optical, or electronic technicians will need a degree in electronics from a vocational or community college. Additional courses in optics and mechanics are highly desirable. More and more programs are being offered specifically for laser technicians.

Even though having a degree is important, experience plays a key role in securing a desirable position. If you are a student, it is valuable to assist your instructors in the laboratories of your scientific courses. If you are employed, it is valuable to seek out work from your present employer that is challenging and that constantly exposes you to new and different experiences. The greater your depth of involvement in a project or a process, the more you will learn and the more valuable you will be to your present as well as future employers.

In Conclusion

If lasers had been invented in 1920, when Einstein first theoretically described stimulated emission, imagine where we might be today! Since the laser was, in fact, invented in 1960, use this time frame to project where we will be in the year 2025.

What could conceivably carry Captain Kirk from the surface of a planet up to the orbiting *Enterprise*? A laser beam, of course. We are still confronted with the slight problem of converting his body to energy, beaming him up, and reconstructing him, and what if we lost part of the signal in between? Using a more realistic means of "travel," however, we can expect to be able to send a three-dimensional image of Captain Kirk to another place and return a three-dimensional image of the occupants of that place to him. In this way a meeting can be carried on just as if they were together. Imagine attending a conference thousands of kilometers away without having to physically go there.

How about a time in the future when doctors rarely have to open our bodies to perform surgery? A fiber-optic catheter and sensors could be directed to the afflicted area, from which a three-dimensional image could be sent back to guide the doctor's work. The doctor could send another fiber-optic catheter to carry the necessary laser energy to the sight of surgery to remove unwanted tissue or weld tissue together.

How about having the entire contents of the Library of Congress available at the touch of a fingertip on your personal computer? It would be sent to you over fiber-optic cables from a large-scale optical-disk storage system. A doctor could slip an optical disk into his personal computer and have access to almost everything known in his field of medicine. The same accessibility to knowledge could apply to scientists, lawyers, engineers, managers, architects, bankers, and anyone who needs large quantities of information.

Imagine one huge machine sitting in the middle of a factory. A worker sits at the computer console in front of it. He enters into the computer the type of part to be manufactured, its size, and the material that should be used to make it. The machine goes to work. Robotic arms move the material into place; microcomputers control the cutting lasers to form the component parts precisely; other robots pull the pieces together and move the welding lasers along the required seams; a laser monitor observes the welding process and controls the laser power to obtain a reliable weld. The finished parts come flying out one side of the machine.

Or imagine an enormous electric generation plant, driven

by the world's largest laser, where frozen hydrogen pellets are heated up to the temperature of the center of the sun, producing enough energy to supply a thousand cities with electricity. With this much electric power available, most people would drive electric cars, eliminating a large part of the use of oil and gas, the major source of air pollution in our cities. In this way the laser will help us to use fusion energy to power and clean up our cities instead of using it to destroy them.

To no small extent, laser development will have an impact on the course of major technical developments in the future. You, because of your fascination with and interest in lasers, could play an important role.

References

MONTHLY PROFESSIONAL LASER MAGAZINES

Laser Focus: The Magazine of Electro-optics Technology
Advanced Technology Group
PennWell Publishing Co.
Littleton, MA 01460

Lasers & Applications
Editorial Headquarters
23717 Hawthorne Blvd. Suite 306
Torrance, CA 90505

Fiber Optics and Communications
167 Corey Road, Suite 111
Brookline, MA 02146

LASER-RELATED PROFESSIONAL ORGANIZATIONS

American Institute of Physics, Inc. (AIP)
335 East 45th St.
New York, NY 10017

Optical Society of America (OSA)
1816 Jefferson Pl. NW
Washington, DC 20036

Lasers and Electro-Optics Society
Institute of Electrical and Electronics Engineers (IEEE)
345 East 47th St.
New York, NY 10017

The International Society for Optical Engineering (SPIE)
P.O. Box 10
Bellingham, WA 98227-0010

Electronics Industries Association
2001 Eye St., NW
Washington, DC 20016

Society for Optical and Quantum Electronics
P.O. Box 245
McLean, VA 22101

SHORT COURSES ON LASERS AND ELECTRO-OPTICS

Laser Institute of America (LIA)
5151 Monroe St., Suite 118W
Toledo, OH 43623

Engineering Technology, Inc.
P.O. Box 8859
Waco, TX 76714

MAJOR ANNUAL LASER CONFERENCES

Conference on Lasers and Electro-optics (CLEO)
IEEE and OSA

International Quantum Electronics Conference (IQEC)
IEEE and OSA

SPIE—Various conferences and short courses

International Conference on Lasers
Society for Optical and Quantum Electronics

International Congress on Applications of Lasers and Electro-optics (ICALEO)
Laser Institute of America

GENERAL BOOKS ON LASERS

Hallmark, Clayton L. *Lasers: The Light Fantastic.* TAB Books, 1984.

Hecht, Jeff and Dick Teresi. *Laser: Supertool of the 1980s.* Ticknor and Fields, 1984.

Maurer, Allen. *Lasers: Light Wave of the Future.* Arco Publishing, 1983.

Muncheryan, Hyrand M. *Principles and Practice of Laser Technology*. TAB Books, 1983.

Index

A

Absorption spectroscopy, 111
Acupuncture, lasers used in, 166
Animation and display system, 184–185
Antiballistic missile system, use in, 100–103
Antisatellite weapons, developing, 180–182
Applications of lasers, generally 6, 106–188
  See also specific subjects
Argon-ion lasers, 99–100
  medical applications, 152
Arterial medicine, application, 159–160
Arthritis, treatment with, 166
Atmospheric optical communications, 171–173
  military uses, 172–173
Atomic elements in lasers, 4–5
Axial-flowing gas CO_2 lasers, 75–76

B

Basic ingredients of lasers, 3–4
  roles of, 5
Basov, Dr. N.G., role of, 17
  Nobel prize awarded to, 17
Battlefield simulations, use of lasers in, 182–183
Birefringence, 41
  birefringent crystals, 41

C

Cancer treatment, use in, 154, 160–161
Carbon dioxide gas dynamic CO_2 lasers, 73–80
  axial flowing gas CO_2 lasers, 75–76
  sealed-tube CO_2 lasers, 75
  transverse-flow CO_2 lasers, 76–77
  transversely excited atmospheric (TEA) CO_2 lasers, 79–80
  wave guide CO_2 lasers, 78–79
Careers in laser technology, 189–196
Cataracts, use in treating, 158
Chemical lasers, 65–66
  output, 65
  space based defense systems, use for, 66
  testing, 181
  types, 65–66
Chromium, use of, 16–17
Coherence
  coherence length, 34
  coherence time, 34
  concepts of, 32
  phase differences, 32–33
  spatial coherence, 33
  temporal coherence, 34
  types, 34
  wave fronts, 32
Collage synthesis, use in, 165–166
Collimated light, effect, 9
Collision broadening, 49

Common characteristics of lasers, summary of, 104
Common types, 69–105
  See also specific subjects
Communications, applicability to, 166–173
Communications technology, use in, 12
Computer engineering, 131–137
Copper vapor lasers (CVL), 100–102
Coronaries, applicability of treatment in re, 159–160

D

Definitions, 5
Dentistry, applicability to, 165–166
Dermatology, use in, 164–165
  collagen synthesis and, 165
Descriptions of lasers, 3
Diabetes, treatment of complications of, 154–155
Doppler effect, 49
Double heterojunction laser, 65
Dye lasers
  application, 89
  described, 86–93
  limitations of, 89
  medical application, 89
  operation, 88
  optical pumping, need for, 87
  uses, 19

E

Education in laser technology, 189–196
  disciplines to follow, 191–192
Electron energy levels, 43–47
  excitation, effect, 46
  properties, 45
  quantized model of the atom, 44
Energy as unit of measure, 29

Energy production sources, use as to, 111–116
Engineering applications, 125–137
Entertainment, application, 183–187
Etalon: optical device, 54
Excimer lasers, 20, 90–93
  application, 93, 153, 158
  availability, 93
  described, 90
  gas laser, as, 90
  medical application, 153, 158
  research, use in, 93

F

Fiber optics
  communications, use in, 167–171
    capabilities, 169
    development, 167–168
    limitations on, 170–171
  military use, 172–175
Foot problems, use in treating, 164
Focused high-power continuous lasers, use, 10
Free-electron lasers (FEL), 20
  advantages, 67–68
  described, 67–68
  research, suitability for, 68
Frequency as unit of measure, 25–27

G

Gain as medium of amplification, 47–49
  collision broadening, 49
  Doppler broadening, 49
Gas dynamic CO_2 lasers, 78
  space based defense, use for, 66–67
    described, 66–67
Gas lasers
  See also specific subjects
  Carbon dioxide lasers (CO_2), 73–80

Gallium arsenide diode lasers, 95–99
  types, 96
Glaucoma, treatment by laser, 155–158
Gynecological uses, 162–164
  cervical growths, 162
  menstrual problems, 162

H

Helium-cadmium lasers (HECD), 100
Helium-neon lasers (HeNe), 69–73
  application, 71–73
  computer printers, use in, 72
  described, 69–73
  inexpensive operation, 71
  stability, 71
  uses, 71–73
Historical knowledge of lasers, 14–21
Holography and holograms, 20, 116–122, 186–187
  described, 116
  effect, 11–12
  interferometry, 116, 122
  limitations, 121
  wave fronts, 116–119

I

Image generation, 185
Industrial applications, 137–150
Industrial process and lasers
  monitoring, inspection, and control, 147–150
Industries where careers available, 189–190
Inertial confinement fusion, 111–116
Interference patterns, production of, 34–38
  interferometer, use, 36–37
  wave mixing, 36–37

Interferometer, use, 36–37
Invention of laser
  Basov, Dr. N.G., role of, 17
    Nobel prize awarded to, 17
  dye lasers, 19
  excimer laser, 20
  first laser, 17
  free-electron laser (FEL), 20
  gas ions, 19–20
  Gordon Gould, role of, 17–18
  historically, 14–21
  noble gas ions, 19–20
  Prokhorov, Dr. A.M., role of, 17
    Nobel prize awarded to, 17
  ruby laser, 17
  Soviet Union, role of, 15, 17
  Townes, Dr. Charles, role of, 15–18
    Nobel prize awarded to, 17
  United States, role of, 15

K

Knowledge needed to understand lasers, 22–42
  birefringence, 41
    birefringence crystals, 41
  coherence, 32–34
    See also Coherence
  electron energy levels, 43–47
    excitation, effect, 46
    properties, 45
    quantized model of the atom, 44
  exceptions to general workings, 61–68
  fundamental concepts, 43–68
  gain as measure of amplification, 47–49
    collision broadening, 49
    Doppler broadening, 49
  interference patterns, production of, 34–38
    interferometer, use, 36
    wave mixing, 34–35

measurement, units of, 23–34
  energy, 29
  frequency, 25–27
  linewidth (bandwidth), 27–29
  power, 31–32
  time duration, 29–31
  wave lengths, 23–25
mode locking, 55–58
  dye laser, use, 56
modes of oscillation, 52–54
modulation, 41, 58–60
  acousto-optic, 58–59
  electro-optic, 59–60
  multiplexing, effect, 42
multimode operation, 54
optical resonators (cavities),
    49–51
  common types, 49–51
oscillation modes, 52–54
phase locking, 55–58
  dye lasers, use, 56
polarization, 39–41
  Brewster angles, 41
  Brewster windows, 41
  polarizers, 40
  unpolarized light, 41
Q-switching (Q-spoiling), 55
refraction, index of, 38–39
single mode operation, 54
summary, 60–61
total internal reflection, 38
transverse electromagnetic
    modes (TEM), 53

L

Laser light displays, 183–187
Laser computer printers, 135–137
  laser scanners, 135
Linewidth (bandwidth) as unit of
    measure, 27–29
Low-power lasers, medical uses,
    10–11

M

Magnetic confinement fusion,
    111–112
Maiman, Dr. Theodore, role of, 17
  first laser built by, 17
Manufacturing, application to,
    137–150
  cutting and drilling, 142–143
  industrial process monitoring,
    inspection, and control,
    147–150
  marking, 145
  materials processing, 138–140
  semi-conductor processing,
    146–147
  soldering, 145
  surface treatment, 144–145
  welding, 140–142
Measurement, units of, 23–34
  energy, 29
  frequency, 25–27
  linewidth (bandwidth), 27–29
  power, 31–32
  time duration, 29–31
  wave lengths, 23–25
Medical applications, 150–166
  acupuncture, 166
  advantages of, 150
  argon lasers, use, 152
  arterial medicine, 159–160
  arthritis, 166
  cancer treatment, 154, 160–161
  cataracts, treatment, 158
  collagen synthesis, 165–166
  common lasers and, advantages
    of, 150
  coronary medicine, 159–160
  dentistry, 165–166
  dermatology, 164–165
    collagen synthesis, 165
    polarized light affecting wound
      healing, 165
  diabetes complications,
    treatment, 154–155
  excimer laser, 153, 158

foot problems, 164
glaucoma, treatment, 155–158
gynecological problems,
    162–164
  cervical growths, 162
  menstrual problems, 162
low-power lasers, use, 10–11
myopia, treatment, 158
neurosurgery, 161–162
oncology, 160–161
ophthalmological problems, 151,
    154–159
photocoagulation and, 152
podiatry, 164
retina, treatment, 150, 155–159
tumors, treatment, 154, 160–161
Military applications, 172–183
Mode locking (phase locking),
    55–58
  dye lasers, use, 56
Modes of oscillation, 52–54
Modulation of the beam, 41, 58–60
  acousto-logic, 58–59
  electro-optic, 59–60
Monochromaticity of lasers, 9
  value of, 9
Multimode operation, 54
Multiplexing, 42
Myopia, treatment, 158

N

Neurosurgery, application to,
    161–162
Nitrogen lasers, 103
Noble gas ions, use, 19
  types, 20
Nonlinear effects, 21
Nonlinear optics, 122–125
  heterodyning, 122
  nonlinear interactions, 124–125
NOVA laser, 112–116
  uses, 112–116

O

Oncology, use in treatment, 154,
    160–161
Operation of lasers. See Knowledge
    needed to understand lasers
Ophthalmological uses, 151,
    154–159
Optical data storage of
    computerized information,
    131–135
Optical devices: Etalon, 54
Optical resonators (Cavities),
    49–51
  common types, 49–51
Optics and precision measurement,
    125–131
Oscillation, modes of, 52–54

P

Phase locking (mode locking),
    55–58
  dye lasers, use, 56
Photocoagulation, benefits of, 152
Podiatry, application to, 164
Polarization, 39–41
  Brewster angles, 41
  Brewster windows, 41
  nonpolarized light, 41
  polarizers, 40
Power as unit of measure, 31–32
Printers, computer, 135–137
  laser scanners, 135
Projection systems based on, 185
Prokhorov, Dr. A.M., role of, 17
  Nobel Prize awarded, 17

Q

Q-switching (Q-spoiling), 54–55
Quantized model of the atom, 44
Quasi-stable states, effect, 4

R

Refraction index, 38–39
Research and development,
    application of lasers and,
    107–125
    See also specific subjects
Retina, treatment, 150, 154–159
Ring laser gyroscope, use, 128–131
Ruby laser, 16–17, 102

S

Scanners, 135
Science and technology, use in,
    6–13
    See also Technology
Scientific knowledge needed to
    understand lasers, 22–42
    See also Knowledge needed to
    understand lasers
Scientific notation, 22–23
Scientific principles, understanding,
    21 *et seq.*
Sealed-tube CO_2 lasers, 75
Semiconductor diode lasers, 61–65
    bandgaps, 63
    described, 61
    double heterojunction laser, 65
    energy bands, 62
    P-N junction, effect, 63
    similarities to other lasers, 62
Semiconductor lasers, 94–98
    described, 94–95
    gallium arsenide diode lasers,
    95–99
    types, 95
    uses, 95
Single-mode operation, 54
Soviet Union, role as to
    development of, 15, 17
Space-based defense systems, use
    of lasers for, 66
Space telescope, use in, 125–128
Special properties of laser light,
    6–9
Special things lasers can do, 6–13

Spectroscopy, 20
    absorption spectroscopy, 110
    applicability, 9
    applications of, 107–111
    biological systems, application to,
    111
    defined, 107–108
    precision measurement, 110
    types, 107
    use of lasers, 107–111
    value of, 109
Steel welding, use for, 10
Strain patterns, revealing by use of
    lasers, 11
Strategic weapons, developing,
    180–182

T

Technology
    careers, 189–196
    education and, 189–196
    science, use in, 6–13
    technicians, requirements for,
    190
Time duration as unit of measure,
    29–31
Total internal reflection, 38
Townes, Dr. Charles, 15–18
    Nobel prize awarded to, 17
Transverse electromagnetic modes
    (TEM), 53
Transverse-flow CO_2 lasers, 76–77
Transversely excited atmospheric
    (TEA) CO_2 lasers, 79–80
Tumors, treatment, 154, 160–161

U

Understanding lasers, what must
    know, 22–24
    See also Knowledge needed to
    understand lasers
Unique characteristics of laser
    light, 6–13
U.S. Air Force, use by, 172–179

U.S. Army, application of by,
     172–175, 178–183
U.S., historical role, 15–16
U.S. Navy, application by, 172–173
Units of measure, 23–34
    energy, 29
    frequency, 25–27
    linewidth (bandwidth), 27–29
    power, 31–32
    time duration, 29–31
    wavelengths, 23–25
Usefulness of lasers, 6

Uses, generally, 6–13, 106–188
    See also specific subjects

V

Vibronic lasers, 102–103

W

Waveguide CO_2 lasers, 78–79
Wavelengths as unit of measure,
     23–25
Weapons, use in creating, 172–183

Credits

Credits for the color plates are as follows:

Page 1, *top left*	Photograph courtesy of Advanced Kinetics, Costa Mesa, California
Page 1, *top right*	Official United States Navy photograph
Page 1, *bottom left*	Photograph courtesy of AT&T Bell Laboratories, Murray Hill, N. J.
Page 1, *bottom right*	Photograph courtesy of Coherent, Inc., Palo Alto, California
Page 2, *top left*	Photograph courtesy of Singer Kearfott Division, Little Falls, N. J.
Page 2, *top right*	Photograph courtesy of Optical Storage International, Santa Clara, California
Page 2, *bottom left* and *right*	Photographs by Ms. Debbie Rousch, R.N., courtesy of Dr. Harold W. Brumley, Austin, Texas
Page 3, *top left* and *right*	Photographs by Dwayne Rolf, ophthalmic technician, courtesy of Drs. M. Coleman Driver and Lyle D. Koen, Austin, Texas
Page 3, *bottom left* and *right*	Photographs by Dwayne Rolf, ophthalmic technician, courtesy of Drs. M. Coleman Driver and Lyle D. Koen, Austin, Texas
Page 4, *top left* and *right*	Photographs courtesy of Perkin-Elmer Corporation, Danbury Connecticut
Page 4, *bottom left*	LASERIUM photograph, courtesy of Laser Images, Inc., Van Nuys, California
Page 4, *bottom right*	Photograph courtesy of LaserMedia, Inc., Los Angeles, California